Biology Made Easy

An Illustrated Study Guide For Students To Easily Learn Cellular & Molecular Biology

Published by NEDU LLC

Written by the creators at

Disclaimer:

Although the author and publisher have made every effort to ensure that the information in this book was correct at press time, the author and publisher do not assume and hereby disclaim any liability to any party for any loss, damage, or disruption caused by errors or omissions, whether such errors or omissions result from negligence, accident, or any other cause.

This book is not intended as a substitute for the medical advice of physicians. The reader should regularly consult a physician in matters relating to their health, and particularly with respect to any symptoms that may require diagnosis or medical attention.

NCLEX®, NCLEX®-RN, and NCLEX®-PN are registered trademarks of the National Council of State Boards of Nursing, Inc. They hold no affiliation with this product.

Some images within this book are either royalty-free images, used under license from their respective copyright holders, or images that are in the public domain.

Illustrations created by our wonderful team. The content was written by the creators of NurseEdu.com. Published by NEDU LLC.

For bulk orders in paperback, please contact: Support@NurseEdu.com

ISBN: 978-1-952914-06-5

FREE BONUS

FREE Download for Nursing Students

Just Visit:

NurseEdu.com/bonus

TABLE OF CONTENTS

PREFACE

Why study cell biology when a single cell is just 1/30,000[th] of the human body? Does what takes place within a single cell really matter? Maybe you would think differently about one small cell if that cell made a simple genetic mistake one day and, a few months later, had transformed into one million cells as a cancerous lump inside your body.

You would think differently about the inner workings of that cell if you had a child with a mitochondrial defect that led to problems producing cellular energy. Your child would have mental defects and muscle weakness for life — all because their cells weren't working.

Even if none of these examples are true for you, you might still find the study of cell biology to be fun and interesting, especially if it is explained to you in an engaging way. That's what we'll do in this book by using plenty of illustrations and analogies so that you can understand what's happening in each cell of your body.

A single cell is a microcosm of you in your entirety. Each one is a kind of miniature factory with the job of taking its genetic information and using it to make proteins that perform all sorts of biochemical and structural functions in your body. Not one of your cells operates in a vacuum. Every cell has a means of identifying itself to other cells and of communication with other parts of your body, both near and far. This is how what happens in each cell becomes important to who you are as a whole person. As you'll see, even single cells like bacteria have means of communicating with one another.

In this book, we will cover all aspects of cell biology to show how well-coordinated the cell system is. Cells are bound by membranes. In fact, part of what defines a *living thing* is that it has a membrane. This is why most biologists do not recognize viruses, which lack these membranes, as living things. You'll see how amazing cell membranes are; they are not just boundaries for the cells. They are very selective about what gets into and goes out of the cell. In that sense, each cell membrane is a very good sentry, providing a gateway so the cell environment is well-regulated.

We will also talk about genetics. This, of course, is the entirety of the DNA inside each cell. Did you know that each cell of your body has an identical set of chromosomes, which are the structures that package your genetic material? How could this be if each cell is different? How does your skin cell make the pigment in your skin while your stomach cells make stomach acid if the genetic "blueprint" is the same? You'll soon see how this ingenious system works.

One of the main jobs of each cell is to metabolize the food you eat and use the oxygen you breathe to create energy. This energy is used to provide fuel for the biochemical reactions that run every single process inside the cell. In this book, we will talk about how nutrients like sugar and fat get taken into the cell, are churned up in complex ways, and become energy used for fuel in order to drive chemical reactions.

The cell does many things — both for itself and for the purposes of making products used by other cells. We will cover each of these things in detail until you get a clear picture of everything that happens inside each cell. By the time you finish this book, you will be intimately familiar with the cell and what it does. Through fun facts and analogies, you will see how all cellular processes are interconnected and what happens if any cell process goes wrong.

In the end, the hope is that you will view the cell differently and realize why everything you do to stay healthy affects each and every cell of your body.

DIVISION ONE:
INTRODUCTION TO BIOLOGY AND YOUR CELLS

Before we get too far into the particulars of cell biology, you should know what biology is all about and what it means to study it. There is a lot more to biology than you likely realize. Sure, there is human biology, which in important to us as people, but what about other living things like plants, fungi, and even bacteria?

SECTION ONE:
THE BASICS OF CELL BIOLOGY

You can't dig too deeply into cell biology without knowing how it really fits into the study of the sciences and biology itself. The study of biology is different from, say, the study of physics, but you will come to see how all of these things are interrelated. If you've studied physics before, you may have heard about *energy*, *work*, and *thermodynamics*. You can be sure that, even though these terms seem unrelated to biology, no part of a biological system functions without these important basics of physics at work at the same time.

By the end of this section, you should be able to say what biology really is, what defines living things, and how scientists have decided to organize them. You should also see why the life forms on earth today look almost nothing like life on earth three to four billion years ago. You will see what parts of biology are encompassed in the study of cell biology in particular. Finally, we will talk about genetics and how we have come to see it as an extremely important part of what defines living things.

CHAPTER 1:
WHAT IS BIOLOGY?

Let's start with the study of biology. How did this idea of "studying biology" start, and what did we know about biology thousands of years ago compared to now? What is a living thing and what isn't? What do we know about cell biology and genetics that helps us understand complex ideas like genetic engineering and human diseases?

The Study of Biology in History

While the word *biology* comes from the ancient Greek term for "the study of life," no one used this term until about 1800 when several scientists, including Thomas Beddoes and Jean-Baptiste Lamarck, used this word for the first time. Of course, this doesn't mean that this is when the study of biology began. Mankind has been looking into the study of animals and plants since ancient times, although it was often called "natural history" or "natural philosophy" instead of biology.

Back in the Middle Ages, biology was sometimes called "natural theology." If that sounds a bit religious to you, you would be right about that. Before the Renaissance (which started in the 15th and 16th centuries CE in Europe), the study of biological systems was conducted either by scientists who were also religious figures or by researchers who knew that, if what they discovered about biology didn't fit with accepted religious principles of the day, it could be bad for the scientists involved. The Church was too powerful and largely dictated which parts of science were true and which weren't.

Even in earlier times, biology was still being studied informally. Farmers living 10,000 years ago knew about plants, including how to get their crops to be more productive by selecting only the seeds from the hardiest plants to grow for the next season. Doctors in ancient times had ideas about the biology of humans that they used to "cure" human diseases. People who kept domesticated animals knew enough about biology to be able to turn "wild animals" into the cows, pigs, and dogs that are mostly domesticated today.

What Ancient Cultures "Knew" about Biology

We know the most about the ancient cultures that studied living things *and* wrote down their findings. This skews things a bit because there could have been cultures from thousands of years ago who knew a lot about life but never documented it, or that had documentation that didn't survive to be looked at today. Animal husbandry and plant domestication during ancient times, for example, was largely an oral tradition passed down from generation to generation without anyone transcribing it.

Let's start with the "cradle of civilization" in Ancient Mesopotamia and the Babylonians, who lived nearly 4,000 years ago. Mesopotamia doesn't exist today; it's now called Iraq. Like other ancient cultures, the concepts of magic and science were essentially the same. They didn't know much about biology as we know it today but understood enough to be able to prevent the spread of diseases.

The ancient Chinese studied herbs, alchemy, and philosophy — often in interwoven ways. As with the Babylonians, biology was studied as it applied to health and medicine. They also seemed to understand that something we now call evolution was happening. The Chinese invented Traditional Chinese Medicine or TCM, which is still an accepted treatment strategy for diseases in modern society.

The Ancient Indians had Ayurveda as their traditional medical philosophy, which was based on "three humors" in living things. They divided life forms into those that came from seeds, moisture, eggs, or the womb. They studied medicinal plants that are still used today.

The Greeks were the first scientists who contributed much of what we know about biology. They were philosophers who asked a lot of scientific questions but only speculated as to the answers. They were big into the idea of humors as the constituents of the body, including blood, black bile, yellow bile, and phlegm. Aristotle studied the classification of living systems based on what he observed about life and the animals he dissected. Others, like Pliny the Elder, did the same thing.

Biology in the Medieval and Renaissance Periods

The Medieval Period or Dark Ages was a real slump time for biology (and for a lot of other scientific studies). There were few scholars who kept studying biology because their lives were at stake if their theories didn't match the Church's belief system.

The scientists of the Renaissance fared better. This was the start of the "scientific revolution," with more researchers reporting on findings ranging from anatomy to natural history. Religion and philosophy fell away in favor of the study of anatomy, physiology, and medicine. There were even artists who studied biology in order to be able to draw with anatomical correctness.

The Last Four Centuries of Biological Study

By the 1600s, biology evolved more closely into the structure we are familiar with today. This was when scientists got serious about organizing and classifying living systems. Carl Linnaeus created his taxonomy texts in 1735, which introduced the system of classifying life that we still use. Scientists gave names to their discoveries and began to theorize about evolution.

Human physiology was always a big topic because it meant better understanding of the human body and how to fix it if something went wrong. Early microscopes were invented so the cells themselves could be studied in ever-increasing magnifications. Scientists started to think small more and more in order to understand the big problems in science and medicine.

By the end of the 19th century, biology wasn't just a subject. It was a super-subject with a lot of different disciplines underneath it. A researcher in biology could be a botanist, cytologist, anatomist, bacteriologist, or physiologist, among others. They started to think of the idea of living systems being machine-like in quality. This way, they could put together microcosms of information in the cell, for example, and extrapolate what they learned to larger biological systems.

The 1800s also brought about germ theory and cell theory as the major ideas of the era. They could see individual cells by using staining techniques and studied them with improved microscopic techniques. They realized that there were some cells that could give rise to whole life forms. They called these *germ cells* and said they were different from *somatic cells* that couldn't create a whole living being when they divided.

They also decided that there was a difference between organic substances and inorganic substances. This happened during the rise of organic chemistry, which is the chemistry of living things. The end of the 19th century marked the first discovery of enzymes used for speeding up biochemical processes in living organisms.

By the 20th century, there were many fields of biology being actively studied, including the biology of the environment, or *ecology*. While the study of genetics was well underway by then, the shift was toward molecular genetics — studying the genetics of cells and DNA rather than an organism's phenotype or its appearance. Massive strides were made in the fields of biochemistry, molecular biology, and all areas of medicine.

Don't forget DNA, the structure of which was first seen in 1952 by Rosalind Franklin using x-ray crystallography. The idea of the DNA molecule being a double helix was floated by James Watson and Francis Crick in 1952, winning them the Nobel Prize. Over time, researchers were rapidly figuring out that DNA was the blueprint for the cell and how DNA passed its message onto messenger RNA. Those, in turn, became proteins in the cell using the genetic code, which is essentially the same process for all life forms.

In the latter part of the 20th century, biology became more technical with the advent of biotechnology, which actually was used before then by brewers and agriculturists. Biotechnology expanded and became an important component of developing drugs, genetically engineering plants and livestock, and refining gene therapy. We will talk later about how things like recombinant DNA technology was used to make vast quantities of specialized drugs like insulin, which in the past had to be harvested from cows and pigs.

What is a Living Thing?

Before you get too far into the study of biology or cell biology in particular, you should ask yourself, "What is a living thing, anyway?" Are all things organic the same as being a living thing or are there criteria we go by to define what's alive and what isn't? As it turns out, most scientists agree on some acceptable criteria for what is life and what isn't. Let's see what these are.

A living thing should have most or all of these characteristics:

- It should have *homeostasis*. Homeostasis is a combination of "sameness" and "resiliency." It means an organism can stay regulated in different environmental conditions. It could be as simple as a one-celled bacterium that maintains the same electrolytes in the cell regardless of the environment or as complex as sweating when you are working out to keep your temperature within normal limits.

- It should be organized into one or more cells. This is a sticking point for some biologists because this rules out viral particles, including viruses and viroids, as being living things, even though they reproduce and have genetic material. This is why you'll hear about viruses "replicating" instead of reproducing because, technically, they don't really reproduce.

- It should have some type of *metabolism*. Metabolism is nothing more than changing nutrients into energy and then using other molecules (plus energy) to make new molecules. The processes of *catabolism* (the breaking down of a molecule) and *anabolism* (the building of a molecule) are both metabolic processes that perform opposing actions.

- Growth must be involved somehow. This basically means that you need to see more anabolism going on than catabolism. The cell must actually grow in some way that is different than just building mass. The different parts of the organism must grow at a roughly equal pace.

- It must be able to adapt to the environment. In a sense, this can relate to adaptation within the same organism or adaption of the species through evolution and heredity. It should be able to adapt to changing external factors as well as to its nutrient sources and availability.

- It should respond to stimuli in some way. A unicellular motile organism can adapt by using whatever motility means it has to avoid a toxic substance in the environment. Plants will turn toward sunlight in what's called *phototropism*.

- It should be able to reproduce in some way. The reproduction could be asexual or sexual. As mentioned, while viruses replicate or have the capabilities of making new copies of themselves, they cannot do this independently and do not replicate through actual mitosis or meiosis, as is true for living organisms.

Most or all of these statements must be true to call something living. There are some biologists who still question whether viruses qualify. Notice that having genetic material is not a specific criterion for being a living thing. This image shows you what this all should look like:

How Living Things are Organized

Living things have been organized in some systematic way since Carl Linnaeus first wrote his manuscript on taxonomy in 1735, called *Systema Naturae*. He was a Swedish botanist and zoologist who first invented what we call *binomial nomenclature,* which helps to define living species in a scientifically reproducible way. Because of Linnaeus, humans are called *Homo sapiens* according to the genus and species that define us.

There are several categories we use in biology to define living things. The study of these categories is called taxonomy. The classifications you might recognize are *genus* and *species*. These are the narrowest categories in this process. There are more. Here's the list from narrowest to broadest:

- Species
- Genus
- Family
- Order
- Class
- Phylum
- Kingdom

There are thousands of known species of living things with more being identified every day. There are even sub-species classifications for organisms technically of the same species that have specific differences requiring refinement of their nomenclature.

The broadest definition before kingdoms, even, is the domain. There are just three domains in life. These are *bacteria*, *archaea*, and *eukaryota*. Both bacteria and archaea are organisms considered to be *prokaryotes*. These are single-celled organisms that are so simple that they have no internal structures besides a ball of genetic material. They divide only through *binary fission*. Archaea used to be considered a type of bacterium. However, after studying their biochemistry, biologists decided they needed their own classification.

Eukaryota is the kingdom that plants and animals as you know them fall into. Eukaryotes are organisms that have internal structures inside their cells. They can be single-celled organisms or multicellular organisms that, while often different from one another, all have the specific characteristics of having organelles inside each cell.

There are just a few kingdoms, however, and you should know what they are. Kingdoms are the broadest category of living organisms, each having key distinctions that make them unique and different from each other. Let's take a look at these.

Back in Aristotle's day, they recognized plants and animals as being the two major kingdoms. In Linnaeus's time, they added minerals as a life form, which made three kingdoms: animal, mineral, and vegetable. After the microscope was invented, they dropped the mineral idea but still had three kingdoms – plants, animals, and protists. *Protists* represented all primitive life forms, regardless of where they really fell in terms of their biology or biochemistry. Protists, or *protista,* were defined as all single-celled organisms.

By 1938, scientists were able to tell the difference between prokaryotes and eukaryotes under the microscope. It led to another rearrangement of kingdoms. There were four in total: *Monera, Protista, Animalia,* and *Plantae.* This led to another idea: How about two empires? There were two of these, known as the Empire Bacteria and the Empire Eukaryota.

Things got more out of hand in the late 1960s when scientists decided to add another kingdom and rearrange things further. This kingdom system was partly based on an organism's nutrition. If you were in the kingdom Plantae, you were an autotroph that used sunlight as an energy source, but it also meant you were most likely multicellular. If you were in the Animalia kingdom, you were also multicellular but derived nutrition through eating. That left *protista* and *fungi* as being separated into their own kingdoms.

By 1977, things expanded even further when scientists determined that bacteria and archaea were, in fact, different from one another. This led to the separate kingdoms of bacteria and archaea. (If you're keeping track, that's six kingdoms.) A few years later, they added *chromista,* which were organisms that had their chloroplasts (which are the organelles involved in photosynthesis) inside a different part of the cell than true plants. These chromists were removed from the kingdom Plantae. They also added a new kingdom called *Archezoa,* but this new kingdom was quickly dropped off the list.

There have been a few proposals and changes over the years but the latest was published in 2015. There are now seven kingdoms. Bacteria and Archaea are both prokaryotic organisms. The other five are all *eukaryotic organisms* and include *Protozoa* (think amoeba), *Chromista* (like brown algae), *Plantae, Fungi,* and *Animalia.*

To Sum Things Up

Biology has been studied loosely for thousands of years. Theories from back in the day were not altogether accurate, nor were they similar to what we now know. The "four humors" notion fell by the wayside and so did the miasma theory of disease causation. Scientists were able to think on a micro level and, by linking cells and their inner structures, were able to understand cellular physiology, which is basically how living systems function.

By now, you should know that when you are studying biology, you are essentially studying the essence of life. As you have learned, there are certain characteristics that define what a living thing should be able to do. Viruses have genetic material and replicate but they do not have a metabolism or grow in any appreciable way. If you think of viruses as being "assembled," which is basically what happens when they infect a cell, you will see why they aren't considered life forms.

All of life is categorized in the study of taxonomy. While there have been many changes since Carl Linnaeus first proposed classifying living things, we now have three major domains and seven kingdoms. Among the three domains are two types of prokaryotic cells and eukaryotes, which are further divided into five kingdoms beneath the domain called Eukaryota.

CHAPTER 2:
THE STUDY OF EVOLUTION

You might think that talking about evolution strays a bit from the study of cell biology, but as we humans evolved from single-celled organisms, it is a good idea to look that far back at how life on Earth started. It probably began with the formation of organic molecules, possibly brought in from asteroids or comets. These molecules came together in complex ways to create simple life forms. In this section, we will talk about how we came to learn about evolution and how biologists think it all went down from the early years on our planet until today.

The Study of Evolution Throughout History

When we think of evolution, we usually think of the famous experiments of Charles Darwin, who studied finches and other animals on the Galapagos Islands. Darwin was able to study an isolated part of the world, where it was easier to analyze different species that had undergone a relatively rapid rate of evolution on small islands. He was able to document how different features of these animals had adapted to the environment over successive generations within the same species.

While Charles Darwin is famous for his work, he was certainly not the first to study evolution. The ancient Greeks got into the game early when Anaximander, an early astronomer and geologist, decided that life started on Earth in a wet environment and gradually evolved into living in a drier place. He felt that mankind was not always the same at the beginning of time. Erroneously, he believed our ancestors were fish and felt that an ancestor to humans must have existed once with features that did not involve being an infant. He largely believed that because infants are helpless, we could not always have existed with this state as part of human existence.

Anaximander and other Greeks of the era (about 450 BCE) set the tone for evolutionary beliefs for many centuries. By the middle of the 1700s, there was a scientist known as Comte de Buffon, who thought that four-legged creatures were perfect and rose above all species, but that out of all of these types of animals, there must have been a common ancestor.

The idea of natural selection or "the survival of the fittest" dates back to Ancient Greece, but it was modified through the years until the time of Erasmus Darwin, who was Charles Darwin's grandfather. Erasmus Darwin thought that natural selection existed but only among the males of a species and that females were only the womb-source of the offspring. He thought that fathers contributed some kind of "filament" that made an embryo without genetic influence from the mother.

Jean-Baptiste Lamarck was a scientist in the early 1800s. He had some interesting ideas that were mostly wrong. He thought that if you used a muscle or body part often (like weight-lifting to build your muscles), you could pass that trait to your offspring. The only thing he got partially right was that evolution happened slowly over time. Mostly, we know that evolution is gradual but that sudden mutations can cause a jump in the evolutionary process.

Researchers in the 1800s put together what farmers who kept animals already knew, and applied these concepts to evolution. Edward Blyth, for example, examined artificial selection, which occurs in animal husbandry when farmers select the quality traits in the domesticated animals they keep, allowing for variation and an evolution of their animals' characteristics over time. Blyth didn't believe you could develop a new species with natural selection; instead, he thought it simply perfected a given species.

Fun Factoid: What do you think "survival of the fittest" really means? Does it mean being stronger, living longer, or having some other trait? Actually, it is some other feature that determines fitness. In evolution and natural selection, the fitter organism is one that produces the greatest number of offspring (that also must reproduce in their lifetime). You could be technically weaker, but if you produced more offspring, more of your kind would be around for the next generation, and you would be considered fitter than a physically stronger organism that couldn't procreate.

Darwin and his colleague, Alfred Wallace, first wrote about natural selection and evolution in 1958. A year later, he wrote his famous work, *On the Origin of Species*, in which he said unequivocally that natural selection could create brand new species. His work was systematic and complete in talking about evolution in these terms, but it wasn't until the 1930s that most scientists accepted his theory.

The study of evolution also did not end after Darwin. He was not particularly interested in studying human evolution because it was pretty controversial at the time. Soon afterward, however, scientists uncovered Java man in the 1890s. This discovery led to the idea that there was an intermediate type of "man" between the Neanderthals and the people of today.

In the 20th century, several things came together to create what we now call *the modern synthesis*. This was a concept that brought together Mendelian genetics, genetic variation, and natural selection. Scientists started to think of evolution in terms of populations, including how certain genes will rise or fall in the population depending on a number of factors. They decided that a species was any population of organisms that could breed together and create offspring that were themselves fertile.

Fun Factoid: Why are horses and donkeys of a different species from one another? They can certainly breed and when they do, a mule is created. Mules have physical differences that distinguish them from horses and donkeys, but these are no more different than if you crossed a St. Bernard with a dachshund to get a dog with mixed features. The difference is that mules are sterile and can't make their own offspring. This defines donkeys and horses as belonging to separate species altogether.

Molecular genetics became understood in the mid-20th century so that biologists understood the structure of DNA and the genetic code. This led to idea of the *molecular clock*, where biologists could tell which species were closer to one another, evolutionarily speaking, by looking at proteins or DNA sequences in order to see if there were similarities. The more similar the structure of the protein or genes, the closer together two species were. This idea is used today to decide whether two organisms are near to or far from one another on the evolutionary tree.

What is Natural Selection?

We touched on natural selection earlier as an important concept in evolution; in fact, it is what evolution is based on. There are always slight differences in a given population of organisms. Some differences do not affect much of anything, while others add to the reproductive success of certain organisms but not others.

If, for example, you were a bird that had a longer and thinner beak so you could reach grubs better than another bird, you would eat better and would (at least theoretically) produce more offspring — who would more likely than not also have longer and thinner beaks. Over time, longer and thinner-beaked birds would predominate over the other birds. Genetically speaking, you would say that the frequency of the gene or genes for this type of beak would drift toward those genes that code for the more advantageous beak. This is called *genetic drift.*

Natural selection also depends on a bit of *selection pressure.* For example, if there was plenty of food and other unlimited resources, there would be no advantage for any of the organisms. You would need to have some sort of pressure on the population, such as competition for food or environmental extremes, that would make some organisms fitter than others.

As animals or other organisms adapt to their environment, there will be some that simply do it better than others. This adaptation will show up as having some characteristic that leads to the animal having an advantage over others in its same niche. You should also know that the animals competing for resources need to be in the same niche. If one animal lived in the swamp but another lived on land, the niches are different, so they really don't compete with one another.

Whatever adaptation the organism has over another must be inherited so that it shows up in the offspring. If this isn't the case, evolution just can't happen from generation to generation. You can see how this could be taken to the extreme so that if the genetic variation between generations gets too big, a species could emerge that wouldn't be able to reproduce with others of their former species.

This sometimes happens if part of a population migrates to a different habitat. As the generations passed in two separate populations with two unique habitats, there is a chance that there could be enough of a drift in genes that the two populations could no longer mate. This process would have led to the creation of two species out of what used to be just one.

Certain circumstances need to take place for natural selection to work. If you have these, the chances of evolution and speciation (the creation of new species) is greater:

1. Phenotypic variation — There is a variety of phenotypes in a single species. (Think of the dog species and the wide variation in breeds.)
2. Unequal reproductive success — There are differences in the number of offspring an animal/plant can produce, depending on the circumstances.
3. Inheritance — Traits get passed on to offspring in lasting, genetic ways.
4. Trait-fitness association — Certain traits affect an organism's fitness or ability to survive in the environment.

You can see that natural selection and evolution depend on inheritance, some visible differences in the population, selection pressure that leads to unequal reproductive success, and the addition of some kind of trait that contributes to the fitness of the organism in some way.

Fun Factoid: In the early industrial age, the levels of pollution and soot in the air suddenly became much greater. In Manchester, England, they noticed that the normally light-colored peppered moth was largely replaced by a much darker version. This was because the darker ones escaped predators better and had an advantage in the soot-filled environment. Once the Clean Air Act of 1956 was passed, the number of light-colored peppered moths returned to what it once was. This is an example of evolution happening over a short period of time.

There are different types of natural selection, as you can see by this image:

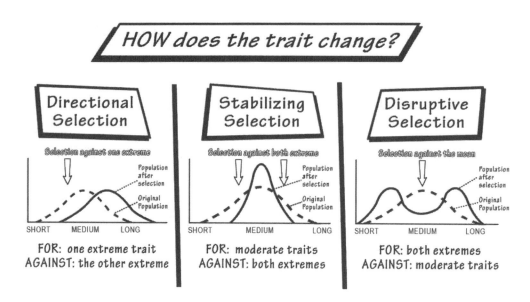

With directional selection, one extreme is favored over all others, leading to more of the extreme organisms in the population. Think of giraffes and their long necks. Their necks might not have been so long in the past, but as it became easier for long-necked giraffes to get food than their shorter-necked cousins, the evolutionary trend was toward the extreme, long-necked version of giraffes.

In stabilizing selection, the "medium" organisms do better than those at either extreme on the spectrum. This means there will be fewer extremes over time and more of the organisms with the same medium appearance. In disruptive selection, there will be more than one extreme that is favored over the medium types. This type of selection has the best chance of leading to two different species developing over time — one at one extreme and the other at the other extreme.

Life as it Evolved from the Beginning on Earth

Some of what we believe about the creation of life on Earth is speculation because there are no fossil records for organisms like bacteria or any other kind of organism without a skeleton. Still, there is a lot we know about early life on this planet that makes sense.

Our best guess is that the Earth itself was formed over tens of millions of years, arising out of dust and gas left over after the sun was created. The Earth was certainly much hotter in the beginning and gradually cooled as it grew in size because of the natural tendency of a heavy celestial body to draw in more and more material over time.

We study the Earth using the geological time scale that measures things in millions and billions of years. The term *mya* means "millions of years ago." The Earth by itself is probably 4.5 billion years old at this time. The atmosphere initially came from volcanic gases; it initially did not have much oxygen. Water likely came to Earth from asteroids or comets. The geologic timeline is long, going from a hot and inhospitable world for us and the organisms of today (largely with methane and CO2 in the air) to an oxygenated, cooler planet where the species we know can exist and thrive.

The first evidence of life on Earth dates back to about 3.5 billion years ago. The Earth's crust had barely formed. Scientists have found fossils from the first microbes in Australia and Greenland that date back that far.

The first organisms that used sunlight for energy in a process called *photosynthesis* probably started making oxygen for our atmosphere beginning about 2.4 to 3.2 billion years ago. Life did not become multicellular until about 580 million years ago, which led to a period called the "Cambrian explosion." It was during that time that there were many more life forms on Earth with the greatest diversity there had ever been. About 99 percent of species that existed during that age are now extinct, even though there are presently as many as 14 million different species on Earth.

The first life forms were prokaryotes. The process of creating life out of organic molecules is called *abiogenesis.* It took as long as two billion more years to have eukaryotic organisms and multicellular creatures on Earth. The first multicellular organisms were probably fungi and/or plants. About 500 million years ago, there was more complex life forms, including birds and mammals, although it is likely that several mass extinctions occurred prior to that point.

There were probably three atmospheres over time on Earth. The first was largely made from hydrogen and helium, which are the same gases coming from the nebula the Earth came from. The solar wind blew this away, giving rise to an atmosphere of water vapor, carbon dioxide gas, methane, and nitrous oxide. These are greenhouse gases, meaning that they heat up the surface of the planet. After photosynthetic organisms arrived, there was much more oxygen. This is not a greenhouse gas, so the temperatures stabilized.

The last common ancestor of all life in our current age were prokaryotes that took in nutrients from the environment and used fermentation to make energy. These early cells didn't require oxygen and may have lived off of methane. As photosynthetic organisms grew in number, there were more developed cells that did use oxygen as a source of aerobic metabolism.

Fun Factoid: Aerobic respiration or aerobic metabolism is just like the aerobic exercise we do today. Aerobic exercise is the kind you do that involves using a lot of oxygen. Because oxygen is readily available, you can go for a long time with this type of metabolism (as long as you have fuel to drive your muscles). Anaerobic respiration is like anaerobic exercise (sprinting or weight-lifting). It doesn't last long in humans because not a huge amount of energy is generated and the resources to drive this type of activity will quickly run out.

The first cells on Earth were very primitive. Some biologists think that those early cells did not have the DNA we currently have but instead used RNA (ribonucleic acid) as their genetic material. They believe this because RNA is easier to make than DNA. If RNA got trapped inside a sphere of lipids, this would meet the definition of a cell. This time period in Earth's lifespan was called the "RNA world." Nowadays, only viruses use RNA as their major genetic material.

Eukaryotes, as we have already mentioned, are more complex with many internal structures. As you will read about soon, two of these are the chloroplasts in plants and mitochondria (in almost all eukaryotic cells). These two organelles are unique because biologists believe that they were once free-living prokaryotic organisms that were engulfed by a larger cell, where they stayed as parts of eukaryotic cells. This process, called *endosymbiosis,* will be further explained when we talk about these organelles in detail.

Multicellular organisms probably weren't found on Earth until about 1.7 billion years ago. We now know that even single-celled organisms talk to one another, so the next step would have been to get these cells to stay together and start to divide their resources. Some cells performed certain activities while other cells performed others.

What About Us? The Evolution of Mankind

We (modern humans) first appeared on this planet about 315,000 years ago as *Homo sapiens*. We were not the first humanoid on Earth; the Neanderthals were here as of 430,000 years ago. You may not know, though, that there were more intermediate humanoids besides these two. For instance, there were also the Denisovans, *Homo heidelbergensis*, and *Homo rhodesiensis*.

Even before humanoids, there was probably a common ancestor that gave rise to modern monkeys, apes, and related primates. Some type of primate lived on Earth as of 65 million years ago. No one knows what the original primate was or what it looked like, but the Plesiadapis lived in North America and the Archicebus lived in China, among others.

The earliest monkeys, called "Old World Monkeys," date back to 20 million years ago. Apes came from these original monkeys about 13 million years ago. They likely lived as far north as Austria at the time. Out of these monkeys came the gibbons, orangutans, gorillas, chimpanzees, and humans.

Fun Factoid: Who among modern primates can we call our closest cousins? Researchers have looked closely at the cell biology and genetics of humans and other primates and have determined the following: The chimpanzees are our closest relative. This is followed (in order) by the gorillas, orangutans, and the gibbons. Further proof has determined that humans parted from our chimpanzee cousins about four to seven million years ago. Hello, cousins!

To Sum Things Up

The study of evolution tells us that life started out with single-celled organisms and that throughout evolutionary time, organisms became more complex and multicellular. There were once a great many different species of organisms on Earth, but biologists think that 99 percent of these have since gone extinct.

Evolution on a small scale is based on natural selection or on *selection pressure*. Selection pressure is when there are differences in the phenotype or characteristics of a population of organisms plus environmental issues that make some characteristics more fit than others. You now know that "fitness" basically means that some organisms have more offspring, which means there will be more of their offspring available for the next generation.

Evolution is both small scale and large scale. We talked about peppered moths that changed their appearance in response to local pollution and changed them again when the pollution disappeared. Evolution on a large scale involves the application of selection pressure over a long period of time that leads to different species being created and evolving independently from others.

CHAPTER 3:
WHAT IS CELL BIOLOGY?

Now that you know a little bit about biology and evolution, let's look at the specific topic at hand in this guide: cell biology. Cell biology as a science did not exist since ancient times. This is because no one could see cells and it would have been impossible to study anything you couldn't see with the naked eye. It wasn't until microscopes were invented that scientists could start researching this field.

Robert Hooke was the inventor of the term *cell* in 1665, when he saw cork under the microscope. He thought they looked interesting and a lot like the *cellula* or tiny rooms that monks lived in. While Hooke was credited with first seeing cells, he really didn't see much because cork cells are basically dead cell skeletons.

Anton van Leeuwenhoek got a little closer to seeing cells about nine years later. He looked at a cell of the algae called Spirogyra and was able to see some of the cells' interior structures. It is likely that he was able to see bacteria as well. Because these were living cells, some of their structures couldn't easily be seen because some type of staining of the cell, which is best done with dead cells, would have been required.

Two ingenious scientists, Matthias Schleiden and Theodor Schwann, got together over coffee in 1836 to compare notes on the animal and plant cells they each had studied. Schleiden studied plant cells and Schwann studied animal cells. They decided that there were a lot of similarities between the two types of cells and determined that cells must be the building blocks of all types of living things. They were a little off, though, because they believed at the time that cells formed spontaneously, like crystals miraculously come from supersaturated liquids. Rudolph Virchow cleared things up by saying that cells only come from existing cells.

All of this research ultimately led to modern cell theory, which is how we have come to see cells today. According to this theory, the following characteristics of cells are true:

- All life forms are made from cells.
- The functional and structural unit of all living things is the cell.
- There is no such thing as spontaneous generation of cells, which means they must come from preexisting cells.
- Cells have hereditary information in them that is passed from generation to generation through the processes of cell division.
- All cells in all life forms have similar compositions.
- The flow of energy in living systems happens within the cells themselves.

Cell biology became a hopping topic in the 1950s as it became easier to grow cells in cultures and use various techniques to manipulate them. When you think of cells in cultures, you probably think of bacterial cultures like those performed to check for infections. You can also grow cells in tissue cultures, which involve eukaryotic cells.

The very first tissue culture that survived and was able to grow continuously involved cervical cancer cells taken from a woman by the name of Henrietta Lacks in 1951. While other cultures did not survive long, Henrietta's cells continued to grow. These were called HeLa cells and they started a revolution in cell research.

Fun Factoid: Henrietta Lacks was a poor Black woman who was dying from cervical cancer in 1951. Her cancer cells were harvested without her knowledge for research purposes. While the woman herself is long deceased, her cells live on in laboratories all over the world. Taking her cells without her permission or compensation sparked an ethical debate about the responsibility of scientists conducting research.

Cells in a Nutshell

The study of cell biology extends to looking at both prokaryotic cells and eukaryotic cells. Prokaryotic cells are really simple. This figure shows you what a typical prokaryotic cell looks like:

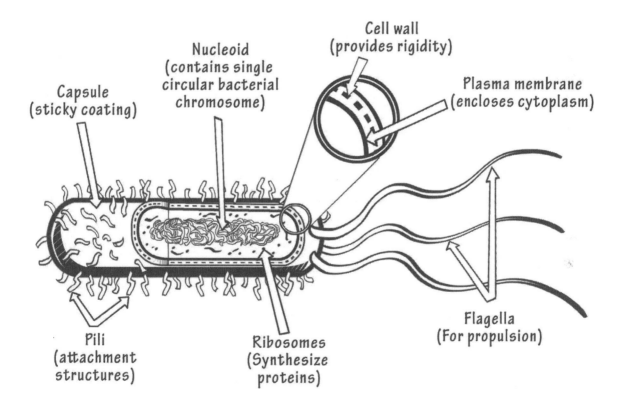

The inside of most prokaryotic cells is fairly basic, but there are some interesting features. Without internal structures (organelles), they have only a liquid interior plus a nucleoid, which is a ball of DNA (genetic material) that isn't bound by any membranes. They also have ribosomes that are close to the nucleoid so that proteins can be made.

The fascinating thing about many prokaryotes is that they have other structures like a cell membrane surrounded by a cell wall. In some bacterial organisms, outside of that is a capsule. Pili and flagella are attached outside the cell surface and used for propulsion.

Eukaryotic cells are significantly more complex than prokaryotes. They have organelles, which are membrane-bound structures inside each cell. Each organelle type has its own job to do in order to make the "cell factory" function correctly. We will talk a lot about how this works later. This figure shows you what a typical eukaryotic cell looks like:

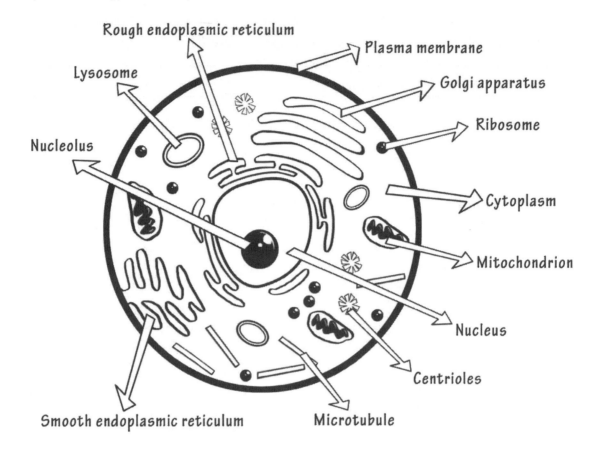

As you can see, there are a lot of different organelles inside eukaryotic cells. Plants and animals are different in a lot of ways but, internally, they are very similar. Plant cells have chloroplasts for photosynthesis, cell walls to help maintain their structures, and water vacuoles to help them maintain turgor pressure when the environment is drier than optimal. Animal cells do not have cell walls but most fungal organisms do. So, depending on what kind of cell you are talking about, there are noticeable differences you'd see under a microscope.

Cell Features

As you study cell biology, you will look much more carefully at the features of different cells. Regardless of the cell type you are studying, there are some similarities that are worth mentioning now.

For example, all cells have metabolism. Metabolism starts with some type of nutrient or nutrients and the ability to process these to make energy. You might think that if a cell takes in a molecule of sugar like glucose, for instance, it would produce some carbohydrate structure within the cell. This isn't how it works, though. Instead, that glucose molecule would make energy so that it could be used to make a carbohydrate structure out of other molecules in the cell.

Cells have different ways of making *energy*. When we say "energy," we are mostly talking about a single molecule called ATP, or adenosine triphosphate. It has three phosphate molecules as part of its structure, and when the molecule breaks up to release one of these phosphate entities, energy is given off that drives certain biochemical reactions in the cell. This is how it works:

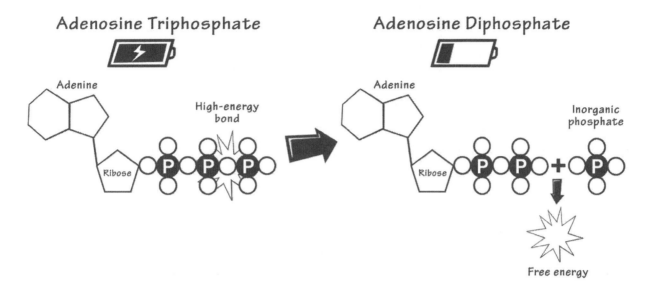

All cells have some type of early process called *glycolysis* that is used to create energy. It involves a biochemical pathway that is essentially the same in all cell forms. It takes a molecule of glucose sugar and breaks it down somewhat to make energy, plus some type of byproduct. In the organisms that make beer, called *Saccharomyces cerevisiae*, the organism will break down sugar with ethanol (alcohol) as a byproduct. Other organisms will have different end products after the glycolysis reaction series. When these organisms do this, it is called *fermentation*.

Glycolysis actually doesn't make very much ATP energy, so multicellular organisms and others that use oxygen will carry forth the process of glycolysis, using oxygen in order to eventually turn a molecule of glucose into carbon dioxide and water. This is called "aerobic respiration" and involves more biochemical pathways. You will later see how a great deal of ATP energy can be created out of just a single molecule of glucose.

Cells also have the ability to communicate with one another. In multicellular organisms, there are three kinds of signaling methods. These include the following:

- Autocrine signaling — A cell sends out a signal that comes back to cause a reaction in itself.
- Paracrine signaling — A cell sends out a signal out that communicates with cells nearby. The signal diffuses outward and travels into cells in its own tissue or in nearby tissues.
- Endocrine signaling — A cell sends out a long-distance signal. In humans, the signal is called a *hormone*. It travels to sites in the body through the bloodstream to act on other cells.

Forms of Chemicals Signaling

Cells also have a certain lifespan called the *cell cycle*. Cells that are not germ cells (which lead to new and unique offspring) will have a specific cycle of growth, duplication of their DNA, and mitosis, which leads to two identical daughter cells after a process called *cytokinesis* occurs. We will talk extensively about this cell cycle so that you can see how a cell goes from being one cell to two cells via a multistep process. Cells will generally die as well. When this happens in a controlled fashion, it is called *apoptosis*.

To Sum Things Up

Cell biology didn't exist as a discipline until microscopes were invented and cells could actually be visualized under higher power than could be seen with the naked eye. Once it was determined that cells had similar features, regardless of the type of cell involved, the cell theory was developed that indicated cells are the functional unit of all living things. Scientists discovered cells are derived only from the division of other cells.

Some cell features include the ability to have metabolism and the ability to grow and divide. Cell metabolism is different depending on the cell type, but the various biochemical pathways are very similar, regardless of the kind of cell. In the same way, almost all cells have the ability to grow and divide in a cycle that is also similar in eukaryotes.

Cells have a specific cycle that starts with growth of the cell contents and proceeds through *DNA replication*, *mitosis*, and *cytokinesis*, which separates a cell into two identical daughter cells. Germ cells are not the same and will instead make gametes so that the male and female gametes can come together to make unique offspring.

CHAPTER 4:
GENETICS AND OUR GENETIC BLUEPRINTS

The study of genetics on a large and small scale is important to cell biology because any cell with daughter cells will pass on genetic information. In sexual reproduction, two separate genders of plants or animals will make gametes that combine to make one or more offspring as a genetic combination of each parent. In this section, we will talk about how genetics has been researched over the years and what we now know about how the process of genetics works.

Genetics is basically studying heredity, such as why you look somewhat like your mother and somewhat like your father. Heredity happens in all organisms but in different ways. In bacteria, for example, the organisms divide through a process called *binary fission*, which gives rise to identical offspring. This does not allow for any genetic diversity, although there can be mutations of the DNA in bacteria that later affect the offspring.

Higher-order animals will engage in sexual or asexual reproduction. Sexual reproduction gives an advantage to the species because it increases the genetic diversity of the offspring. With genetic diversity, the species has a better chance of survival in situations where the environment could become unpredictable.

We now know that genetics involves the genetic material (DNA) in the cell. Only in viruses will the genetic material be RNA instead. We will talk about what DNA looks like extensively later, but for now you should know that it is a long-chain molecule of nucleotides. The nucleotides are a shortened version of an alphabet, while the DNA molecule takes this alphabet in certain orders to create a whole message.

Any message that codes for a single product (usually a protein) is called a *gene.* In eukaryotes, the genes are aligned in larger segments called *chromosomes*. The number of chromosomes is 46 (23 pairs) in humans, but this number will be different for each organism. Bacteria have just one chromosome that isn't linear but is instead in a circular shape.

Fun Factoid: You might think that the number of chromosomes reflects how complex an organism is, but this simply isn't true. Plants, for example, tend to have a lot of chromosomes despite their relative simplicity. Ferns have a total of 1,260 chromosomes compared to us humans, who just have 46 chromosomes. Red ants and gorillas outdo us as well, with 48 chromosomes each.

What do genes do? Each gene on a chromosome is a segment of the structure that codes for at least one protein. Most code for just a single protein, while some will code for more than one. The type of proteins depends on the way the messenger RNA gets cut up after it is made. It's this piece of RNA that gets turned into a specific protein.

History of Genetic Study

When you think of genetic research, you might think of Gregor Mendel and his famous experiments on pea plants. It turns out that there were those who looked at genetics to some extent back in ancient times. Scholars like Aristotle, Pythagoras, and Hippocrates knew nothing about cell biology or genes, but they understood that traits were passed from one generation to the next. Animal husbandry experts of the era (and before that) knew how to select the "best" of their herd to breed for the next generations.

Hippocrates invented an idea he called *pangenesis.* According to this theory, the whole organism contributed traits that got passed onto offspring. He was wrong, however, that male sperm mixed with female menstrual blood in order to make any new offspring.

Epicurus lived in Ancient Greece and also had ideas about genetics or how traits got passed on to the next generation. He observed that both mothers and fathers added something to their offspring and knew that there were dominant traits and recessive traits that could be passed on to any potential offspring. He invented the idea of "sperm atoms" that were sorted and segregated in similar ways as Mendel discovered later with his pea plant traits.

There were other ancient thinkers in various parts of the world who studied genetics. The East Indians felt that, besides parts from the mother and father, the pregnant mother's diet and something called the "fetal soul" helped to determine what a child looked like. Arab doctors understood that hemophilia was a genetic disease.

Anaxagoras in Ancient Greece believed in the "preformation theory" of genetics. The idea behind this theory is that the egg or perhaps the sperm cell was already in a preformed state but only needed to grow over time. This was the predominant theory until Mendel's time. Before this, however, scholars believed in *epigenesis*, which posited that organs grew and developed as the embryo itself grew. This idea was abandoned for many centuries in favor of preformation theory only to be revived as more researchers understood genetics and embryology.

Gregor Mendel and His Laws

Gregor Mendel was a Moravian Monk who was gifted five acres of land for his "experimental garden." He decided to study pea plants because he thought there were a lot of different traits to look at and that they were easier and faster to breed than animals (which was an earlier idea of his). He did his research between 1856 and 1865, studying seven different traits of pea plants. This image shows you the traits he studied:

Seed		Flower	Pod		Stem	
Form	Cotyledon	Color	Form	Color	Place	Size
Round	Yellow	White	Full	Green	Axial Pods	Tall
Wrinkled	Green	Violet	Constricted	Yellow	Terminal Pods	Short
1	2	3	4	5	6	7

Mendel tested almost 28,000 plants and wrote a paper called "Experiments on Plant Hybridization." He reported his findings to the Natural History Society of Brno in Moravia; however, most scientists did not recognize its value until the first part of the 20th century. Nowadays, we understand how important Mendel's work was and have labeled him the "father of modern genetics."

Fun Factoid: Mendel really wanted to study mice and their hereditary patterns but, as he was a practicing monk, he needed the approval of his superiors. Because studying mice involved animal sex, the higher-ups in the church were not happy with this. He switched to studying plants instead. He was able to study the hawkweed plants as well, but these experiments did not work because there is asexual reproduction involved, which skews what the offspring look like.

At the time, genetics were thought to be blended, which meant that any offspring were simply a blend of the parents. This does not make sense to us now because it would lead to offspring that would all be averages of the parents, which isn't generally the case.

Mendel did not understand genes or chromosomes, which hadn't been identified yet. He called the inherited characteristics "factors" and based his theories on five simple rules:

- All plant characteristics he studied were both unitary and discrete. This meant that there were either purple or white flowers that could be easily identified.
- There were different choices of genetic characteristics. One is inherited from each parent. We now call each of these choices *alleles*, which are different forms of the same gene.
- One allele is usually dominant over the other. We now know this to be true only some of the time.
- Gametes (sperm and egg, for example) are randomly generated but with equal statistical frequency. In other words, the "factor" leading to a white flower is equally transmitted to the offspring as the "factor" leading to a purple flower.

- The factors or genes are not linked to one another. This means that they are independently inherited. We now know that this is only partially true and that genes located close to one another are more likely to be inherited as a single unit.

Here's what Mendel did. He initially started with purebred purple flowers and purebred white flowers. He made them by crossing these plants repeatedly to make sure only the color of flower he wanted was created. These were known as the parent generation.

When he crossed the parent generation, he got all purple flowers, which he called the F1 generation. This was a generation of all-hybrid flowers. The cross of the F1 generation yielded what he called the F2 generation. These were an interesting blend of three-fourths purple and one-fourth white flowers. In order to explain this, he created what's called a Punnett square that would be used to determine which of these traits was dominant and which was recessive. This is what a simple single-trait Punnett square looks like:

We now know that a pea plant could have two white genes (and be white in color), two purple genes (and be purple in color), or a combination of white and purple genes (and also be purple in color). In a sense, the white flower gene is "silent" if the purple flower gene is present. We now refer to having two of the same gene (of either color) as being *homozygous* for the gene. If the plant has two different alleles of the flower color gene, the plant would be called *heterozygous* for flower color.

When we study genes in the present, we don't talk about factors or traits but call the different forms of the same gene *alleles.* Alleles are inherited. Mendel understood that the egg cell was held in the plant and the sperm cell was in the pollen. At the time of fertilization, the alleles combined to make a homozygous or heterozygous offspring plant. The actual gene combination in any organism is its *genotype*. For Mendel, what he saw on the outside was the plant's appearance, or *phenotype*.

Mendel then created his laws of inheritance that have led to what we call *Mendelian inheritance patterns.* The first law was called the *Law of Dominance and Uniformity*. He determined that, since the F1 generation produced all purple plants, the purple trait must be dominant over the white trait. This led to a trait being labeled either *dominant* or *recessive* in nature. All recessive phenotypes must have both recessive genes and no dominant genes to interfere.

Mendel also proposed the *Law of Segregation.* This meant that different factors of the parents were separated or segregated in each gamete. There would be equal numbers of dominant purple alleles and recessive white alleles created from any parent that had both alleles as part of its genotype. We still denote all dominant traits with a letter that is upper-case and all recessive traits with a lower-case letter. For example, if purple is P, then white is p. A purebred purple plant has a PP genotype and a purple phenotype. A hybrid has a Pp genotype and a purple phenotype. The purebred white flower has a pp genotype and a white phenotype.

When you look at the F2 generation, what you see is a 3:1 ratio of purple to white flowers when it comes to the phenotype. With regard to the genotype, however, you would get a 1:2:1 ratio instead (look carefully at the Punnett square to prove this).

Mendel finally settled on his third law, called the *Law of Independent Assortment.* This law is only valid when there is more than one trait inherited at the same time. For example, what if you tried to cross plants with some combination of wrinkled versus smooth pea pods and green versus yellow pea pods? If these are inherited separately, you get what is called a *dihybrid cross*. The ratio you end up with is 9:3:3:1. This makes for many combinations of different phenotypic traits.

According to the *Law of Independent Assortment,* the two traits get assorted separately, giving rise to a variety of different but predictable combinations. Theoretically, you can do this with more than two traits at a time, but the Punnett squares would be very complex.

We now know that independent assortment doesn't always happen. As you will see when we talk about meiosis, if the genes are on the same chromosome and are located near to one another, there will be a greater likelihood that they are inherited together, which really messes up the statistics of what you'll see in the offspring. This idea of two genes getting inherited unexpectedly together is called *genetic linkage.*

After Mendel's work was later combined with other genetic research, they came up with what's called a *Mendelian trait.* This is any trait that can be determined by just one gene that is located at a single locus on the chromosome. Things like obesity and diabetes are rarely Mendelian traits, while some human diseases, such as Tay-Sachs disease and sickle cell anemia, are definitely inherited in a Mendelian pattern.

After Mendel

Mendel's work wasn't really identified as being important to the study of genetics until about 1900. Researchers worked with fruit flies and other non-plant species in order to see how traits were passed on through the generations. Some exceptions to Mendel's work were discovered, which we will touch on in a minute. After DNA was discovered, work in genetics expanded greatly and we came to understand things like alleles, chromosomes, genes, and genomes. The organism's genome, for example, is the entirety of all of the DNA it has in each cell. Genes that caused certain diseases were soon identified, which helped parents decide whether or not to pass a certain gene to their offspring.

Different Types of Genetic Study

There are three major types or branches of genetic study that have evolved since Mendel's time. Mendel studied *classical genetics,* which looks at what can be visibly seen after sexual reproduction has taken place. Others have expanded on his work to look at what is seen when certain genes are inherited.

Classical genetics looks at *diploid organisms*, which are those that have two copies of a gene. Remember that these gene copies are not necessarily identical. For example, the genes for yellow or green pea pods are inherited in a diploid organism but are not identical. These are called *homologous genes.* Homologous genes are those that contribute to the same phenotype but, like unmatched socks, they perform the same function while looking different from one another.

Haploid organisms are basically the gametes — a sperm or egg cell. Each of these cell types has just one copy of each gene or allele. If you look closely at a sperm or egg cell in humans, you will see 23 chromosomes instead of the diploid number of 46 chromosomes. When these cells come together, they create a diploid zygote, which has all 46 chromosomes (one set of 23 chromosomes from each parent).

Another branch of genetic study is called *molecular genetics.* This only came after we discovered the molecular basis of inheritance using knowledge of DNA, genes, and RNA. Genes could be identified using special techniques, and biotechnology companies could identify ways to make recombinant drugs by finding genes that could make these drugs and amplify their activity. Molecular genetic techniques were used to identify the entire human genome.

The third field of genetic research is called *population genetics*. This is closely related to the study of evolution. It looks at how traits add to the ability of an organism to adapt to its environment and how new species are created. By studying things like the frequency of an allele in a population and how it can vary over time, researchers can look at how evolution proceeds on a genetic level.

Different Types of Inheritance Patterns: When Inheritance Isn't Mendelian

So far, we have talked mainly about Mendelian inheritance, which is the kind explained by Mendel's work. It is based on the idea that there are dominant and recessive traits or alleles and that there is one gene for every trait. This leads to diseases that are autosomal dominant or autosomal recessive. An autosomal gene is one that is on one of the 22 truly matching pairs of chromosomes in humans. There is a 23rd pair, which might not be matching, mainly if you are male.

These non-matching chromosomes are called the *sex chromosomes*. In humans and most other species, these will be the X chromosome and the Y chromosome. Females are defined as having two

X chromosomes, with a karyotype called 22, XX. Males have an X and Y chromosome, with a karyotype called 22, XY. Your karyotype is what your chromosomes look like under the microscope.

This is what a typical karyotype looks like for a human male:

Fun Factoid: Not all species of animals have the classic XX versus XY patterns in females and males, respectively. In some insects, the female will have a total set of matched chromosomes, including two of the same sex chromosomes. Males have one of these sex chromosomes but have no unmatched Y chromosome equivalent, so they have one fewer chromosome than the females. Not every organism's sex chromosomes are called X and Y, either. There are those that not only have a ZW sex chromosome system, but the males are ZZ and the females are ZW.

When you look for an autosomal dominant trait or disease in a family, you will be able to create a pedigree. If the disease is autosomal dominant, it will be seen in all generations and in males and females equally. The chances of passing the disease onto your children if you have it is 50 percent. This is because you usually have one dominant allele with the disease and one recesssive allele (without the diseased gene). If you are super unlucky and have two copies of the dominant disease-carrying gene, you will pass it onto each of your children without exception. This is what a pedigree would look like in an autosomal dominant disease or trait condition:

Autosomal Dominant

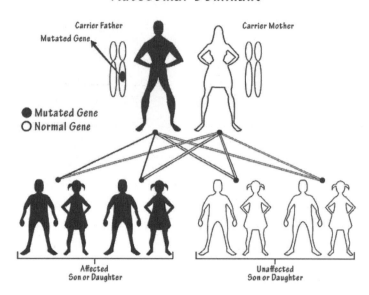

If the dominance is complete, the recessive gene will be completely silent, and you would see no evidence of it unless that person had children who did not have the disease. Remember, there's a 50:50 chance if there is a recessive gene but a 100 percent chance of passing it on if there is no recessive gene present.

There is also such a thing as *incomplete dominance.* This is when the heterozygous person or organism looks somewhat different from the homozygous person or organism. In snapdragons, for example, the red is dominant but, if you cross a red with a white snapdragon, you will get snapdragon offspring that are pink in color. There are other cases where it is subtler so that there won't be as obvious a difference between those that are heterozygous and those that are homozygous for the dominant trait.

Autosomal recessive diseases in humans are common. These are often disorders where some kind of biochemical pathway is involved. In these cases, you would have one normal gene that would make the right enzyme for the pathway and one that would not be functional. If you had two genes for the nonfunctional enzyme, you would be sick with the disease. As long as one was normal, you would not have the disease. The condition called *phenylketonuria* is a common autosomal recessive disease where you don't have all the enzymes needed to metabolize the amino acid called *phenylalanine*.

If you drew a pedigree of this disease, offspring would have it if both parents were carriers or if one had the disease and the other was a carrier. In the more common case of two healthy parents who were carriers, about one-fourth of their children will develop the disease. Another 25 percent would be completely unaffected, while 50 percent would be carriers. This is what it looks like:

Autosomal Recessive Inheritance

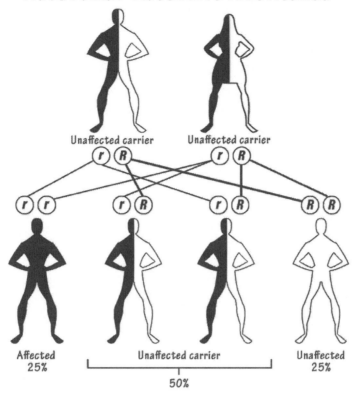

Other diseases and traits will have x-linked inheritance, where the gene for the trait is on the X chromosome. Most of these are recessive traits. Since girls have two of these chromosomes, they will almost always be carriers, while half of all boys in the family would have the disease if the mother was a carrier. No fathers, even if they have the disease, will pass the disease onto their sons (but all of their daughters would be carriers). This is what it looks like:

X-Linked recessive inheritance

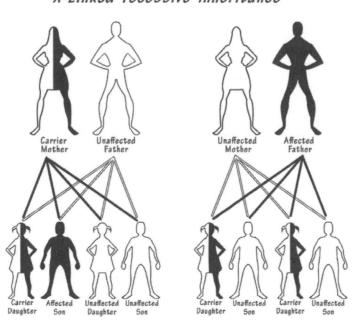

Red-green colorblindness is an x-linked trait. The same is true for most types of hemophilia and Duchenne muscular dystrophy.

Codominant inheritance is seen with blood types. You've likely heard about being a certain blood type, such as O or AB. These designations represent which proteins are in your blood cells. You get one protein from your mother and one from your father; no protein is considered dominant over another. This is what it looks like in red blood cells:

	Group A	Group B	Group AB	Group O
Red Blood cell type	A	B	AB	O
Antibodies in plasma	Anti - B	Anti - A	NONE	Anti - A and Anti - B
Antigens in red blood cell	"A" Antigen	"B" Antigen	"A" and "B" Antigens	NONE

There are also genes called *lethal genes*. These will kill some of the offspring, often before birth. The end result is that the offspring you expect to find with an autosomal dominant trait, for example, isn't in the percentage you would actually see.

One example of this is with yellow mice. Normal mice are brown. If you mate yellow and brown mice, you will see half brown and half yellow offspring. The yellow color is dominant over the brown color. If you take two of the yellow mice and mate them, you will only see two-thirds yellow and one-third brown mice. The reason for this is that any yellow mouse embryo with two yellow color genes (a homozygous yellow dominant mouse) will die before being born, so you wouldn't be able to factor those in the Punnett square the way you'd think. This what it looks like:

	A	A^y
A	Agouti Coat AA	Yellow Coat AA^y
A^y	Yellow Coat AA^y	Dead $A^y A^y$

Fun Factoid: The autosomal dominant human disease called Huntington's is lethal because, if you have the gene, you will develop neurological dysfunction and dementia that results in death. The reason this dominant disease gets passed on at all is that most people will already have had children before they start to develop symptoms. It's rough for those kids because they will know early on that they have a 50:50 chance of getting the disease themselves. There is fortunately now a blood test (but no cure) to see who has it and who doesn't early in life.

Other lethal genes in humans include those for Tay-Sachs disease, Duchenne muscular dystrophy, and cystic fibrosis. These people often don't pass it on themselves because they tend die young, but because these are all recessive diseases, they get passed on through people who are carriers. Some biologists think that each of us has at least one lethal recessive allele in our cells, but because they are so rare, the chances of mating with someone who has the same lethal gene is slim.

Most common traits and even common diseases are not Mendelian because they involve many genes that add up to the total effect. This is called *polygenic inheritance.* Diabetes, heart disease, obesity, and your height are all due to the combination of many genes. Each gene has a small effect on the total picture or phenotype and there are so many of them that all tall people, for example, don't have the same set of "tall genes." There is a continuous averaging of what you'll see in the offspring and the environment often plays some role in whether or not you have the disease or trait. Even though there is averaging happening in the next generation, it is possible for you to be taller than both of your parents — not just an average of the two heights.

To Sum Things Up

Gregor Mendel did a lot to set the stage for what we know about genetics, even though he didn't know anything about genes or DNA. He identified "factors" and developed laws that work well for most single-gene traits and diseases.

Mendel studied *classical genetics*, which involves looking at the appearance of offspring. After things like DNA and chromosomes were understood to play a role in genetics, researchers turned to study *molecular genetics.* Finally, there is *population genetics*, which looks at how traits are carried forth in whole populations.

There is more to inheritance than Mendelian genetics. Some traits are *codominant*, where no gene is dominant over another. There is *x-linked inheritance*, involving the X chromosome-related traits or diseases. There are *lethal genes* that do not allow any offspring to survive to adulthood; some offspring are unable to survive life in the womb and are never born. Most common traits and diseases are *polygenic,* which means that many genes play a role in what the offspring look like.

DIVISION TWO:
THE CHEMISTRY AND PHYSICS OF LIVING CELLS

Undertaking a review of physics and chemistry before digging into cells and what they do seems like a step back but it is a necessary step. Even if you've studied these things before, staying on top of these basic concepts will help you really understand how the cell works. You might think that biology and chemistry are two different things, but that is far from the truth. Your cells follow all the rules of biochemistry, physics, and chemistry.

In the next section, we will talk about chemistry and physics as they apply to cell biology and your cells. It is the same thing as understanding gasoline and automobile oil before studying how they fuel your car. Knowing these simple stepping stones to the big picture of cell biology will help you learn cell biology in an easier way.

SECTION TWO:
THE STRUCTURE OF MATTER:
ATOMS AND MOLECULES IN LIVING SYSTEMS

The chemistry of living things is both similar and different from the chemistry you might have already studied. There are some basic elements common to living things that are not as common in other parts of nature; you should know the properties of each of these elements. You should also be able to understand concepts like *chemical bonds* and the *macromolecules* of living things.

Why get so basic about atoms, elements, and bonding? It's because all of cell biology really is very basic. Cells do not come together miraculously. They follow all of the chemical rules of bonding as well as the physics of atoms and molecules. If you know these things well, a lot of how a cell works will make much more sense to you. Let's forge onward!

CHAPTER 5:
GETTING DOWN WITH ATOMS

You've probably heard of atoms before, right? These are technically the smallest unit of matter—but with a catch. Scientists refer to atoms as being the smallest unit of "ordinary" matter, which doesn't include the many subatomic particles we now know exist. Atoms actually have three major subatomic particles that make them up. These are called *protons*, *neutrons*, and *electrons*. This is what the simplest atom (hydrogen) looks like:

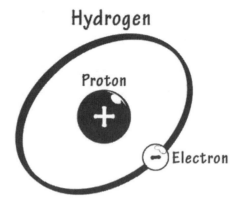

In a hydrogen atom, there is one proton and one electron. While it looks like the electron orbits around the proton, this isn't technically true. Electrons are infinitesimally small and have a negative charge. They are attracted to protons, which are positively charged but much bigger in total mass.

You need to think of electrons not as orbiting like a planet would. Instead, think of the whole environment around a proton as being a cloud, where the electron could be anywhere. Because of their size, charge, and rapidity of movement, they are never in one place at one time. You have to think of them as flitting about in a cloud without any particular orbiting movement.

The center of an atom is called the *nucleus*. It will be positively charged, even if it has some neutrons in it. The nucleus provides an attractive (think magnetic) force on the negatively charged electron. As you will see, while the pull on the electrons is great, atoms form molecules by "sharing" electrons with others.

Any atom that loses an electron altogether will be positively charged (because it has lost a negative charge). Any atom that gains an electron will be negatively charged (because it has added a negative charge). Some atoms are more likely than others to gain or lose electrons. The ability to do this is how molecules are created. The atom that has gained or lost an electron is called an *ion*. This is a basic picture of a lithium ion. It should have three electrons to match its three protons but it doesn't. It has lost an electron, so it is a positively charged Lithium (+1) ion.

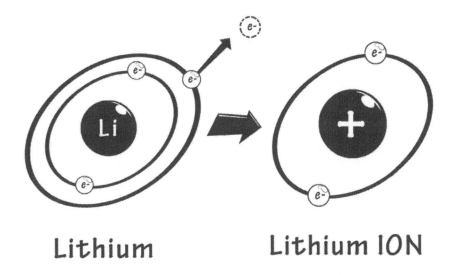

Lithium   Lithium ION

Protons and neutrons have about the same size. The main difference is that neutrons have no charge and just add to the total mass of the atom. Actually, protons and neutrons don't weigh very much either: only about 1.67 x 10^{-24} grams each. Because this is a hard number to write, scientists have decided that a proton or neutron weighs about one atomic mass unit or one *amu*. One amu is also called a *Dalton*. When we talk about the atomic number of an atom, we are talking about its mass in Daltons.

Every atom has its own number of protons, electrons, and neutrons in it. The weight of electrons isn't counted in the mass of the atom, mainly because one electron is only 5.489×10^{-4} atomic mass units. It just doesn't count. When you look at a periodic table of the elements, you'll see the atomic mass in Daltons (or amu units) for each atom. Here's what you'll see on the periodic table for oxygen. Take note of the atomic number and atomic mass on its entry in the table:

Oxygen in the Periodic Table

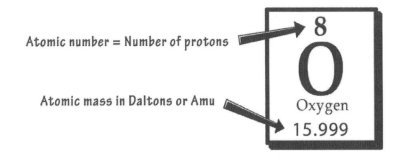

The periodic table shows all the elements. In the upper left-hand corner is the letter "H" for hydrogen. In that box, you'll see the atomic number in the upper left-hand side of the box. This is the number of protons in hydrogen. Below the name, you'll see the atomic mass, which is roughly the same as the atomic number because there are no neutrons to add to its overall mass.

When you go to the upper right-hand side of the periodic table, you'll come to helium or He. This has two protons, which gives it an atomic number of 2. On the other hand, its atomic mass is around 4. This is because it has two protons plus two neutrons, which double the mass. If you subtract the atomic mass from the atomic number, you can figure out how many neutrons are in the atom.

Fun Factoid: Why is it called the periodic table of the elements and why isn't it a nice, neat rectangular shape? The table is conveniently arranged from the smallest to the largest atoms. It is not a neat rectangle because it was designed to line up so that each column, called a period, *has elements with similar properties based on the numbers of protons and electrons. For example, the far-right period or column are the noble gases, which are generally very stable and do not react with any other elements.*

So, what's an *isotope*? An isotope is an atom that has the same number of protons and electrons in the parent atom but with extra neutrons. The total mass of the atom is different from the atom normally seen in nature. Isotopes exist in nature, too, but not to the degree of the parent atom. Isotopes aren't really important in biochemistry, but they are important in things like radioactivity, where the breakdown of a heavier isotope (the loss of one or more neutrons) gives off radioactive energy.

Hydrogen, for example, has three known stable isotopes. One is the parent atom, also called *protium*. It has one proton and one electron. Deuterium is twice as heavy because it has an added neutron on it. Tritium is three times as heavy because it has two added neutrons on it. This is what these look like:

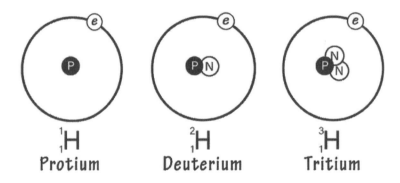

$$^1_1H$$
Protium

$$^2_1H$$
Deuterium

$$^3_1H$$
Tritium

Remember, it is the electromagnetic force or a sort of magnetism that keeps the electron attached to the proton as much as possible. Protons and neutrons stay together because of a type of nuclear force. Nuclear force is actually stronger than electromagnetic force.

When we talk of "splitting an atom," we are talking about breaking the nuclear force of an atom, which causes a neutron to break away from the nucleus. This is such a great force because the broken-off neutron bumps into other atoms, causing a chain reaction of what's called *fission.* If it happens in great numbers, it is explosive.

Fun Factoid: You can see how the power of fission could be really deadly — and it is. In a controlled fashion, fission of heavier elements will cause the nuclear reactions in nuclear power plants. The energy given off when these atoms are "split" is called radioactivity. *In a less-controlled situation, fission can*

Remember how elements in the periodic table have different properties according to the columns they are located in? Well, one of these properties is called *electronegativity*. An atom that is really electronegative doesn't want to give up the electrons it has, while atoms that aren't so electronegative don't mind giving up their electrons as much.

An element that is very electronegative will easily become a negative ion — actually stealing electrons from an atom that is less electronegative. An atom that isn't very electronegative is much more likely to become a positive ion because it will more readily "donate" or give up its electrons. The periodic table is an ingenious way of showing you which elements are more or less electronegative. Take a look at a periodic table. Electronegativity increases from the lower left-hand side to the upper right-hand side.

The most electronegative elements are those on the right side of the periodic table, in particular in the upper right-hand section. These are likely to steal electrons and to form negatively charged ions. When we talk about chemical bonding, this will mean even more to you. For now, just know that there are some elements more likely to make negative ions and others more likely to make positive ions, depending on their characteristics and where they are on the periodic table.

Fun Factoid: You've probably heard of elements like carbon and hydrogen, but what about Berkelium? Yes, it is a real element, but once you learn about it, you might think someone cheated in calling it an element. It has an atomic number of 97, so it has 97 protons in it. It is not found in nature but was created in a laboratory in 1949. Only one gram of it has ever been made. Two of its isotopes are called Bk-250 and Bk-249. Bk-249 has a half-life of 330 days. At some point, it can gain a neutron to make Bk-250, which has a half-life of three hours. All 20 of its isotopes are radioactive (which means they decay spontaneously). Is it cheating to have an element you made in a lab that doesn't last very long?

Electrons: Orbits, Clouds, and Orbitals

It would be nice if all of the electrons in an atom simply orbited the nucleus just as the planets orbit the sun, but it just doesn't work that way. The only similarity is that there is force that keeps these bodies near the larger central structure.

With the solar system, it is the force of gravity that keeps planets from randomly flying away from the sun. In atoms, however, it is the force of electromagnetism that keeps the electrons in place. As we have talked about, however, it is possible to draw electrons away from the nucleus. But in most cases, it takes energy to be able to do this.

If you've studied quantum mechanics at all, you know that all tiny particles are actually physical particles and waves at the same time. This is called *particle-wave duality*. The smaller something is, the more likely it is to behave a lot like a wave.

But wait, what's a particle and what's a wave? How could something be both of these things? First, let's look at what a wave is:

See how it's all spread out? A wave is up sometimes and down at others. It also travels through space, just like light waves travel from the sun to earth at the speed of light. There are slow waves, such as radio waves, and very fast waves, such as x-rays and gamma waves.

When we think of a particle, we don't think of it moving at all. It looks like this:

not here   here   not here

When something is behaving like a particle, it just is. You can see it, and it doesn't move from place to place. This is easy when you think about a rock, for example. You can reach out and touch it without expecting it to jump out of the way. But it doesn't work that way for all particles.

Imagine this: You are trying to catch a puff of dandelion hair floating in the air. You see it and you reach for it only to have the air perturbed enough that it floats away before you can catch it. You try again and the same thing happens. It's awfully hard to catch something so small and lightweight that won't sit still long enough to grasp.

Werner Heisenberg studied this sort of thing. He knew that things with a lot of mass didn't have a lot of momentum to get up and go just when you reach for them. Rocks are heavy. There's no momentum there, so it's not hard to see and touch them. Atoms and electrons are very light, however — especially those teeny electrons. According to Heisenberg, something with almost no mass at all will have a lot of momentum. In fact, they have so much momentum that you can't just grab onto one without it being gone the moment you have your eyes and hand trained on it.

This led Heisenberg to develop his *Heisenberg Uncertainty Principle*, which says that you can't identify the position and momentum of an object at the same time. Momentum is the same as an object's mass multiplied by its velocity or speed. This means that electrons (with almost no mass) are very speedy, so you can't just catch one sitting still without it being gone as soon as you look for it. In this way, electrons are a lot more like waves in their behavior than rocks are.

Fun Factoid: Are you a particle or a wave? According to quantum mechanics, you are both (but not all at the same time)! The larger something is, the more likely it is to behave like classical physics and solid articles. The smaller something is, the more likely it is to behave according to quantum mechanics like a wave. Really, though, you are still a wave in theory some of the time.

Electrons do orbit the atomic nucleus, just not in a nice, neat circle. Because of particle-wave duality and the Heisenberg Uncertainty Principle, atoms have *orbitals* instead. Chemists came up with the idea of orbitals mathematically in order to define where electrons are most likely to be located. They had no way of actually seeing an electron without it disappearing to another location as soon as they looked in a given place for it, so they decided to define where an electron was most probably located.

An orbital, then, is a mathematical concept based on the probability of finding an electron there. With hydrogen, it was easy. There was just one electron and one proton, so the orbital they calculated looked an awful lot like a spherical cloud around the nucleus. They called this the *s orbital* (think "spherical"). This is what it looks like:

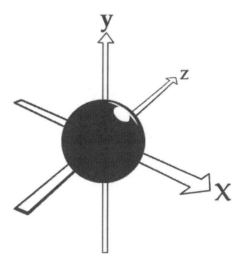

It gets fancier after this. This s orbital is a low-energy state for a hydrogen atom. Without putting energy into it, this is where the electron will most likely hang out. There are other orbitals possible for every atom and element that aren't so simple to imagine. Again, you can't actually see these orbitals; they are based only on mathematical probability.

According to scientists who study these things, only two electrons can fit into one orbital. Electrons have the ability to spin. One will spin one way and the other will spin the other way. They call this a *positive and negative spin.* (Liken it to how negative and positive charges react to create neutrality.) Electrons are balanced when there is one spinning positively and one spinning negatively.

Once an orbital is full, it's happy and it has no real desire to change things. Take helium, for example. It has two protons and two electrons. It also has an s orbital. This is handy because its two electrons can fit in there — as long as they spin opposite to one another. Because this is a "happy" low energy state, helium has no real desire to add or remove any of its electrons. This is why helium is called one of the *noble gases.* Noble gases all share the same quality: their orbitals are filled neatly and the desire to interact with other elements is essentially zero. They are essentially hanging out a "no vacancy" sign.

S orbitals can get bigger as the atom itself gets bigger. You can have a 1s orbital (as in hydrogen and helium), a 2s orbital, and even higher order s orbitals. All of these orbitals are spherical in shape, but they just get bigger. They look a lot like spherical Russian nesting dolls with one outside another, which has another outside of it, and so on. In order for an electron to stay in a higher s orbital (or any higher orbital, for that matter), there needs to be a stronger nuclear charge to keep it there. This means that only larger atoms will have electrons in these higher and bigger orbitals. They have extra protons in the nucleus to add to their electromagnetic ability to hang onto electrons further away from the center.

The next most common orbital in the chemistry of living things is called a p orbital. These are shaped differently. This is what a p orbital looks like:

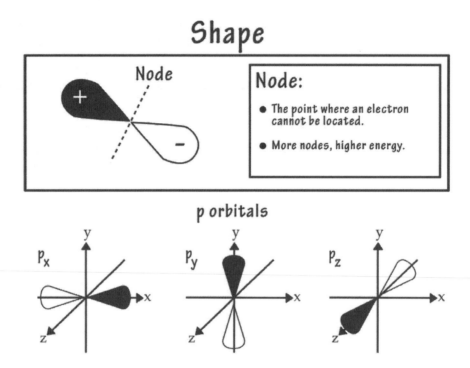

While there is one s orbital per energy level in an atom, there are three p orbitals instead. If you think of a 3D graph as having an x, y, and z axis, there is one p orbital for each axis around the nucleus. The node in the figure refers to any place where the electron has been determined not to ever be (according to probability). As long as an electron is in its lowest possible orbital in terms of the energy it takes to be there, it is happy to stay.

With three p orbitals, there is room for six more electrons. So, for an element like boron, which has five protons and five electrons, there is room for two electrons per s orbital (the 1s and 2s orbitals) for a total of four electrons, plus an extra electron fitting into one of its p orbitals. There is no 1p orbital set but there are three 2p orbitals, called px, py, and pz orbitals.

As you get to bigger and bigger atoms, you will have more orbitals filled up. They don't get filled up randomly but rather according to the energy it takes to keep the electron in the orbital. The less energy required, the better the situation is. It turns out that there is a specific order of orbital energy levels (and it's not what you'd think). The 1s orbital is very low energy, so all atoms have their first two electrons there. Next is the 2s orbital. Two more electrons fit there. After this are the three 2p orbitals (where six electrons can fit). Look at this image to see what this orbital filling strategy looks like.

H (+1) ↑ 1s

He (+2) ↑↓ 1s Filled shell, inert gas

Li (+3) ↑↓ 1s ↑↓ 2s Active!

Be (+4) ↑↓ 1s ↑↓ 2s

B (+5) ↑↓ 1s ↑↓ 2s ↑ 2p — —

C (+6) ↑↓ 1s ↑↓ 2s ↑ ↑ 2p —

N (+7) ↑↓ 1s ↑↓ 2s ↑ ↑ ↑ 2p

O (+8) ↑↓ 1s ↑↓ 2s ↑↓ ↑ ↑ 2p

F (+9) ↑↓ 1s ↑↓ 2s ↑↓ ↑↓ ↑↓ 2p Active!

Ne (+10) ↑↓ 1s ↑↓ 2s ↑↓ ↑↓ ↑↓ 2p

Na (+11) ↑↓ 1s ↑↓ 2s ↑↓ ↑↓ ↑↓ 2p ↑ 3s Active! __ 3p __ __

Mg (+12) ↑↓ 1s ↑↓ 2s ↑↓ ↑↓ ↑↓ 2p ↑↓ 3s __ 3p __ __

After you fill all the 1s, 2s, and 2p orbitals, you go up another energy level to the third level. Then, you start over with 3s and 3p orbitals. You only go to an orbital higher than that if you get past the 3s orbital. Orbitals also come in d orbitals and f orbitals. There are 5 d orbitals and 7 f orbitals — both of which have really odd shapes and neither of which are seen in smaller elements. You have to get up to scandium with 21 electrons to even be able to start filling any d orbitals.

This image shows you the energy levels of the orbitals. They will each fill, two by two, like students in bunk beds until you reach higher and higher energy levels:

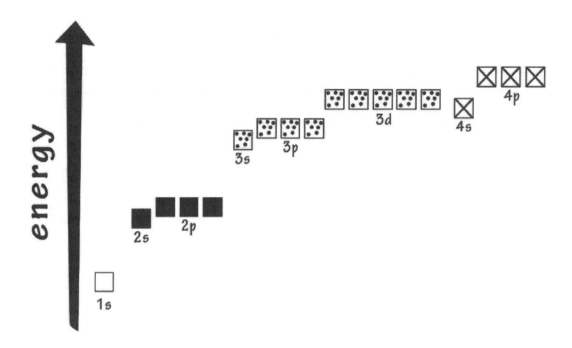

When you look at the periodic table, the dip in the middle starts with scandium. This whole section is called the *d block* because it is in these elements that the d orbitals get filled.

So, what do orbitals have to do with anything, and what can we say about an element just by looking at its orbital filling pattern? Remember the noble gases and how inert they are? This is because their orbitals are filled and there is no room for more electrons. Elements interact or "react" with one another to make molecules by sharing electrons and orbitals. By sharing, the molecule of two separate reactive elements is happier and at a lower energy levels because they both behave like their respective orbitals are filled. It takes less energy to be together than it is to be apart.

Let's take hydrogen, for example. It just isn't possible to see one atom of hydrogen floating around. It has a 1s orbital with just one electron in it when it would be much happier with two electrons filling the whole 1s orbital. So, it gets together with another hydrogen atom (which is also not happy by itself) and they share electrons.

This is a lower energy state with two protons and two electrons sharing their respective 1s orbitals. This is what it looks like:

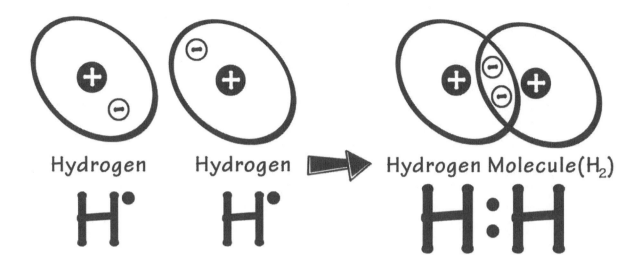

With larger elements and atoms, it gets much more complicated, but it all explains why some elements bond with four possible bonds to another element and why other elements won't bond at all. Neon, for example, is another noble gas. It has 10 electrons. It fills the first four with the 1s and 2 s orbitals; it fills the other six with its three second-level p orbitals. This fills up the second level completely, so it really doesn't have the need to react with any other elements.

What about carbon? It has six electrons and fills up the 1s and 2s orbitals. This leaves the second-level p orbitals. There are two electrons to fill six total spots. That leaves four empty spaces on the second energy level. In order to be "happier," it really wants to fill these up with electrons. So, it shares with other atoms. One stable molecule carbon it can make is called *methane* or CH_4. Why this configuration? Well, each hydrogen atom has one electron to share and the single carbon atom has four spots (one for each electron in a hydrogen atom). You get a nice, stable molecule with all of the second-level orbitals filled in the carbon atom.

When you get to something like fluorine, which has nine electrons, this fills the second-level orbitals almost completely, but it is missing one spot in one of the p orbitals. This makes it especially desperate to grab an electron from anywhere it can. This is the very definition of *electronegativity.* If it comes in contact with a sodium atom, for example, which has an electron in the third level to spare, this is like a match made in heaven. You have one element with an electron to spare at a high energy level plus another that is desperate to get one in order to lower its own energy level. The sodium atom "gives" its electron to fluorine and both are happy.

Atoms and Valence Shells

This idea of the chemical bonding of an atom is called its *valency*. Valency is related to the ability of an atom or element to bind to other elements. The number of bonds an atom forms is called is *valency state*. With carbon atoms, for example, its valency is 4 because it has the ability to make up to four bonds with other atoms.

The outer shell or outermost energy level of an atom is called its *valence shell*. Each of the electrons in the outer valence shell is called a *valence electron*. The number of valence electrons "missing" from an energy level or the number of extra electrons at an energy level determines how many bonds it prefers to form. The closer an element is to either dumping an electron or picking up an electron in order to fill its valence shell, the more desperate it gets to do exactly that.

To Sum Things Up

While we think of atoms as being the smallest units of matter, this isn't strictly true. Atoms have subatomic particles, including *protons*, *electrons*, and *neutrons*. While all elements have some combination of these particles in them, they differ in the exact number. Some elements are tiny while others are huge.

The atomic number of a neutral element is the number of protons or electrons it has. All neutral atoms have the same number of protons and neutrons. Many elements also have neutrons in their central nucleus, adding to their atomic mass but not changing the charge on the nucleus (which has a positive charge). Atoms with uneven numbers of protons and electrons in them are called *ions*. Two atoms of the same atomic number but different atomic masses are called *isotopes* of one another.

Electrons are so tiny and have almost no mass. They behave like waves almost as much as they behave like particles. They flit in random directions around the atom's nucleus and prefer to stay in areas where the energy level is the lowest possible. Instead of orbits (like the planets), electrons settle two-by-two in orbitals, which are mathematical models of where an electron should be based on the notion of staying at a low energy level.

Atoms will bond to another atom depending on what its orbitals look like and how filled up they are. If all the orbitals at a specific energy level are filled, the atom won't want to react to other atoms. If there is room for at least one electron on an energy level, the atom will want to borrow one from another atom in order to fill its valence shell. If a valence shell has a spare electron, on the other hand, it is cheaper (energy-wise) to give that electron up to an atom that wants or needs it.

CHAPTER 6:
HOW CHEMICAL BONDS COMBINE ATOMS

In this chapter, we will talk about how molecules are made by combining two or more atoms. Some atoms don't really do this because they are already happy with their filled valence shells. These are the noble gases we have talked about in previous sections. Other atoms are more or less happy with sharing the electrons they've got, so that as a whole molecule, the total package has less energy in the system than is true of the atoms if they are separate.

A molecule is a group of atoms that are collectively held together by some type of chemical bond. A molecule is, by definition, neutral in charge. You could call an ion (which is electrically charged) a molecule as well, but ions like to combine with other ions to make a neutral molecule. You could call that neutral grouping a *molecule*. Some ions are made from more than one atom. These are called *polyatomic ions*, though some chemists call these "molecules" as well, even though they are electrically charged. The ammonium ion, or NH4+, for example, is a polyatomic ion that you could also call a molecule.

A molecule can be *homonuclear* or *heteronuclear* in nature. A homonuclear molecule would be something like hydrogen gas or oxygen gas, called H2 or O2 gases, respectively. They are termed homonuclear because they are made from two or more of the same atoms. A heteronuclear molecule is one with two or more different atoms in the molecule. There are zillions of these, such as methane (CH4) or water (H2O).

We define molecules by their name and their chemical symbol. You don't need to memorize all of the chemical symbols of the periodic table, but you should know the most common symbols used in the molecules of living things. We will talk more about these molecules in a bit, but some of the more common ones are carbon (C), hydrogen (H), nitrogen (N), oxygen (O), and sulfur (S).

When you write a molecular formula, there are different ways to do this, but the most common way is to use the elemental symbol plus the number of atoms of that element in the molecule. Water, or H2O, can also be written as H_2O. It has two hydrogen atoms and one oxygen atom. This part is easy. It gets harder if you write C6H12O6, which is the chemical symbol for a number of elements, including a molecule of glucose. It's too hard to figure out this way, so instead, you can show it using a few different formats.

Glucose as a molecule can be in a ring shape or in a linear shape. You could write it out as C6H12O6, but that wouldn't define glucose specifically. Instead, you would show its chemical makeup this way:

Glucose

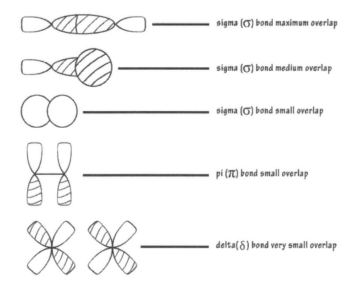

α - D - glucose

β - D - glucose

You can see where it is easy to identify the C, H, and O elements in the linear molecule but, in organic chemistry and biochemistry, there is a shorthand version, often used in ring-forms, where the C is omitted. When you see the ring and the side chains on it, you can assume that the ring is made from carbon atoms except where it specifically says otherwise.

A Brief Look at Bonds and Orbitals

The reason we looked at orbitals earlier was to show where electrons hang out around the nucleus. In order to form a bond with another atom, the electrons (which basically do all the bonding anyway) cannot just get up and leave an orbital in order to make a bond. They have to make their bond while still staying in their orbital lane. This leads to a variety of possible bond types from an orbital perspective.

Two common bonds in biological systems are called *sigma bonds* and *pi bonds*. A sigma bond happens any time two orbitals actually overlap end-on with one another, while pi bonds happen when two lobes overlap side-by-side. They look basically like this:

sigma (σ) bond maximum overlap

sigma (σ) bond medium overlap

sigma (σ) bond small overlap

pi (π) bond small overlap

delta(δ) bond very small overlap

There are really tight sigma bonds and sigma bonds that aren't as tight. Pi bonds are not tight at all because the two orbitals sort of "graze" one another instead of overlapping in an end-on way. In some situations, the whole thing can get really complicated with s orbitals bonding with other s orbitals, s orbitals bonding with p orbitals, or two p orbitals bonding with each other. Some of these pairings look like this:

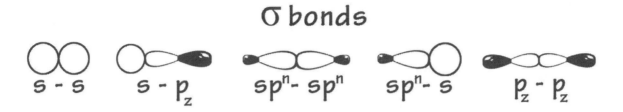

In the above figure, where the bond forms between two atoms, the orbital gets bigger in order to accommodate the bonded/paired/shared electrons between the two nuclei. It can get even messier when you think about double and triple bonds between carbon atoms in organic chemistry. The orbital bonding in ethylene, for example, is a double bond involving a sigma bond and a pi bond that looks like this:

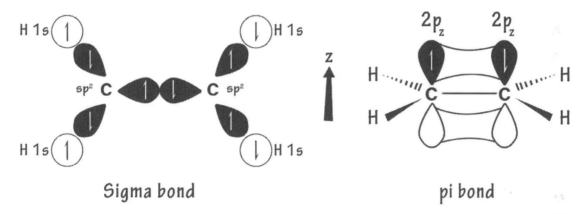

The image shows how a double bond can occur using two separate orbital types. This involves the only available orbitals the carbon atoms have to share their electrons with each other.

Types of Bonds

In chemistry, not all bonds are created equally. There are four main bonding types you'll see in living systems. They differ greatly in how strong they are and depend entirely on which atoms are being connected to one another to make the molecule. Let's look at the different bond types you might see:

In *covalent bonding*, two atoms combine with one another to essentially share the electrons relatively equally. Carbon is a common element in living systems that only engages in covalent bonding. This is because carbon is not terribly electronegative, so it doesn't want or need to completely "steal" an electron. It has two electrons in its valence shell with room for four more. This means that it will share its spare p orbitals with another molecule in order to form a very stable bond.

Covalent bonds in general are extremely stable; this means that it takes a lot of energy to break this type of bond. The sharing of electrons is a trade-off for the atoms involved. Remember, there are two nuclei involved and both are positively charged. Normally, two positively charged nuclei would repel each other. When they share electrons, they must still do this while maintaining their distance. The electrons keep the nuclei together just enough so a bond is formed.

Now, it is a completely equal sharing experience if the two atoms in the bond are identical. The charge on each atom and the size of each atom is the same, so any tug-of-war for the electrons between the two nuclei will always be a draw unless there are other forces in place. This figure shows a covalent bond between two hydrogen atoms:

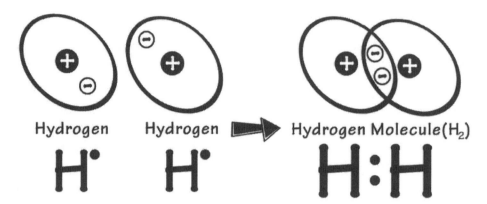

The electrons in the middle are called *shared electrons* or *electron pairs.* There are different ways you can look at these pairs. One of these is called a Lewis dot structure. It allows you to see the bond as well as the electrons not involved in the bond at all. There are some tricks as to how to do this. Take a molecule of carbon dioxide or CO2. The carbon atom will have four valence electrons while oxygen will have six. They will be able to share four of these electrons, forming two double bonds between each of the atoms. The leftover electrons are written as dots and the bonds themselves are written as lines. Each line stands for a single pair of valence electrons. Imagine four electrons sharing the same space between the carbon atom and each oxygen atom. This what it looks like:

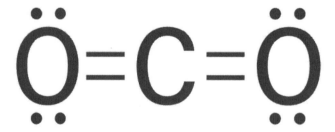

While this structure looks like a nice linear or straight molecule, it isn't that way in three dimensions, as we will talk about soon.

Fun Factoid: The chemist who invented the Lewis Dot Structure was Gilbert Newton Lewis. He lived from 1875 to 1946 and invented this method of identifying bonds based on their electron features. He was nominated for a Nobel Prize 41 times but never won. At the age of 70, he had lunch with a rival chemist (who actually did win a Nobel Prize) and was found dead a few hours later in his lab, which was filled with hydrogen cyanide gas. Many who knew Lewis felt he had become upset by something that had happened at the lunch and committed suicide hours later from cyanide poisoning.

If the two atoms are widely different in size, one of the atoms will have a stronger charge on it and will be greedier for the electrons. There will still be a covalent sharing of electrons, but it won't be 100 percent equal as it would be if the two atoms were identical to one another. The electrons (if you could actually see them) would be pushed further towards the bigger atom with a larger charge. This is what you'll see with water or H2O.

Ionic bonding is the kind of bonding you'll see with salts, in particular. A salt like sodium chloride, or NaCl, involves a sodium (which is technically a metal) and chloride (which is called a *halogen*). Chlorine is very electronegative. This means it is greedy for electrons. Sodium, on the other hand, really wants to get rid of one. Because sodium has what chlorine wants, an ionic bond forms, which is more like a donation from the atom of one electron to the atom of another rather than a shared electron situation. This is what it looks:

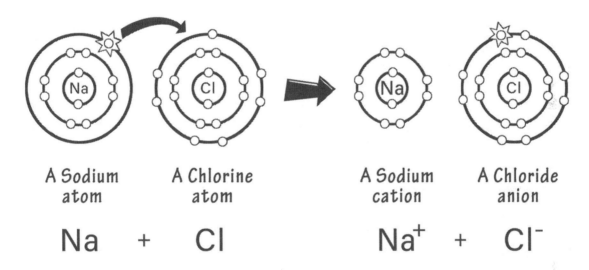

A Sodium atom   A Chlorine atom   A Sodium cation   A Chloride anion

$$Na + Cl \qquad Na^+ + Cl^-$$

Ionic bonding is important in chemistry and biochemistry. The positively charged atom is called the *cation* and the negatively charged atom is called the *anion*. It's an extremely tight bond when sodium chloride or any salt is in its solid form, but not necessarily so in a solution of water.

This type of bond is interesting because in water, the sodium and chloride ions will essentially float in a sea of cations and anions, keeping the charge neutral overall but in an open kind of way. The bond isn't as tight between an individual sodium and chlorine atom. Instead, the electrons will be shared in a collective way with all of the sodium and chlorine atoms in the solution.

Because the electrons are shared among a lot of different atoms, they will move around. The movement of electrons is basically what electricity is all about.

That's why ionic or salt solutions conduct electricity. This is what sodium chloride looks like in a watery solution:

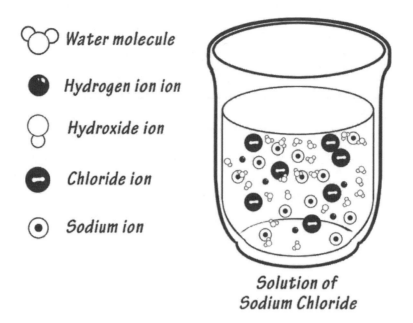

Solution of
Sodium Chloride

Polar Versus Nonpolar Bonding

What is the difference between a polar and a nonpolar bond? A *polar bond* is a lot like having a north and south pole on Earth. You have a magnetic field with the compass pointing north. The north pole has one magnetic charge, and the south pole has the opposite magnetic charge. Polar bonds are like that and, in a very tiny way, these too have a magnetic charge.

The magnetic charge in a polar bond means that one atom has a positive charge and the other atom has a negative charge. It's a tiny difference, not enough to amount to much on a large scale, but very important in chemistry. We call the charge difference between the two atoms a *dipole moment*. Polar bonds have larger dipole moments than nonpolar bonds. When you see a chemist writing a dipole moment on a charge, it looks like this:

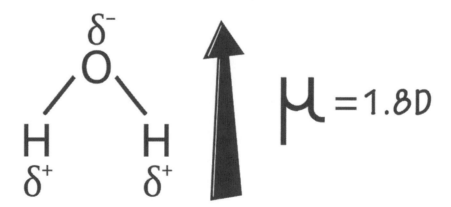

You can think of a dipole moment as being the combination of a magnetic charge difference *plus* distance between the two charges. The delta (δ) symbol stands for partial charge on an atom, while the mu (μ) symbol stands for the actual magnitude or size of the dipole moment as measured in Debye units. One Debye unit is the same as 3.34×10–30Cm3.34×10^{-30} Coulomb meters. A *coulomb* is a unit of electrical charge and *meter* is a unit of distance. The greater the charge difference *and* distance between the atoms, the greater the dipole moment will be.

A good way to think of polar bonds is to go back to the idea of electronegativity. Again, when you see the word *electronegative*, just think "electron greedy" when describing that element. Any time you get a molecule or bond where there is an electron-greedy element and one that isn't so electron greedy, this is the perfect recipe for a polar bond. The end result is this:

Polar Bond = Greater Electronegativity Differences

You can determine the electronegativity of an element and give it a value. The famous scientist Linus Pauling developed an electronegativity scale that ranges from 0.7 to 4.0. The least electronegative elements are cesium and francium, which have an electronegativity score of 0.7, while fluorine is very greedy, having the greatest electronegativity score of 4.0.

A polar covalent bond isn't as dependent on the actual electronegativity of the elements in the bond but the *difference* in the two electronegativities. This table describes the different types of bonds based on this difference:

Predicting Bond Types

Electronegativities help us predict the type of bond:

Electronegativity Difference	Type of Bond	Example
0.00 - 0.40	covalent (nonpolar)	H - H
0.41 - 1.00	covalent (slightly polar)	H - Cl
1.01 - 2.00	covalent (very polar)	H - F
2.00 or Higher	Ionic	Na^+Cl^-

Notice that ionic bonds are in this mix. Ionic bonds are basically polar bonds on steroids. They are so polar, in fact, that the charges are essentially completely split apart. One atom takes the electron nearly entirely, while the other atom gives it up entirely. On the opposite end of the spectrum are nonpolar bonds, which have very little difference in electronegativity between the two atoms. In other words, nonpolar bonds share nicely between the two atoms involved.

Think about oxygen, or O2, gas. Oxygen by itself is fairly polar with a score of 3.5. This means that it certainly could create a polar bond with something, but, when it binds with itself, the electronegativity difference between the two molecules is zero, so it's entirely nonpolar. All bonds between identical atoms will be nonpolar for the same reason.

Common polar molecules are water (H2O), ammonia (NH3), and ethanol (C2H6O). Nonpolar molecules involve almost all hydrocarbon molecules, such as benzene (C6H6), methane (CH4), and the hydrocarbon liquids toluene and gasoline.

In biological systems, there are lots of polar molecules, such as water, salts, and carbohydrates. Fats or lipids, on the other hand, are very nonpolar. In chemistry and biology, like molecules dissolve in like solutions. Polar and nonpolar just do not mix with one another. This is why you can't get oil and water to mix together in a real solution when you shake your vinaigrette. You get bubbles of fat interspersed in water but no real dissolution of fat into water.

When we study membranes of cell structures, you will realize just how important these concepts of polar and nonpolar really are. The outside of biological membranes is polar but the inside, which is basically fatty, is entirely nonpolar. You will not get a polar molecule to slip through a nonpolar membrane. They don't like each other, so it won't happen without a lot of help, biochemically speaking.

In biological systems, when we talk about a polar molecule, we say it is *hydrophilic*, which means "water-loving." These are molecules that like water and dissolve easily in it. On the opposite end of the spectrum are nonpolar or *hydrophobic* water-hating molecules, which will not dissolve in water. The old adage really is true: "Oil and water do not mix." Oil will dissolve in a hydrocarbon solution like gasoline but not in water.

Fun Factoid: Why does detergent cut grease and help clean your oily dishes? A molecule of soap or detergent is long and skinny. It has a polar hydrophilic end and a nonpolar hydrophobic tail. It forms a spherical shape in water with the heads on the outside but has the ability to trap grease in this oil bubble interior. The image looks like this:

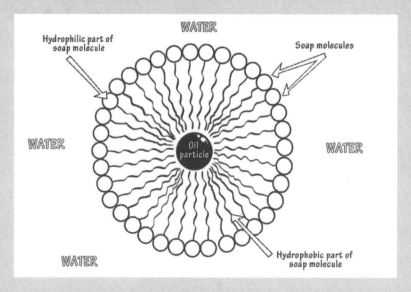

When you then wash the soap off, the oil gets washed away as well since it's trapped inside each bubble of detergent.

In biological systems, *hydrogen bonding* is also very important. Hydrogen bonding isn't like covalent bonding because it is more like an attractant force between two different atoms than it is an actual bond. Nevertheless, it is an important force in biochemistry and in living systems.

Hydrogen bonding involves any molecule that has a hydrogen atom attached to it. Hydrogen, when it is attached to an electronegative atom, like oxygen in a water molecule, will be a bit positively charged. Look again at the image of the water molecule with the partial charge differential between oxygen and hydrogen. This positive charge on hydrogen is tiny but it is enough to attract other molecules. Rather than being a bond between adjacent atoms, hydrogen bonding can be a force between two different molecules instead.

This image shows hydrogen bonding between water and the nitrogen atom in ammonia, or NH3:

The dotted line between the hydrogen and nitrogen atoms (between the two molecules) is the hydrogen bond or force that attracts the two molecules. In the same way, the oxygen in water can form a hydrogen bond between it and the hydrogen atom in ammonia.

Hydrogen bonding is weak but if you add the hydrogen bonds to a beaker of water, for example, it really adds up. That's what makes water so sticky — electrically speaking, that is. Water has such a high cohesiveness among its molecules because of this amazing hydrogen attractive force going on throughout the liquid water.

Hydrogen bonding accounts for a lot of different properties of liquids and even solids. It's because of hydrogen bonding that a molecule will have a higher solubility in water, a higher boiling point, and a higher melting point. This is because it will take more energy to break this hydrogen bond apart. Ice is less dense than liquid water (which is very unusual for most liquid/solid combinations) because hydrogen bonds between solid water molecules help it form a less dense lattice shape as a solid.

Hydrogen bonding is really important in biological systems. In DNA, for example, the molecule involves two separate strands that look like a ladder with rungs. The "rungs" of the ladder involve nitrogenous bases. These rungs are only held together by one thing: hydrogen bonding between the bases. When these hydrogen bonds are broken, the strands can separate so that the DNA can be replicated in the cell.

This is what it looks like:

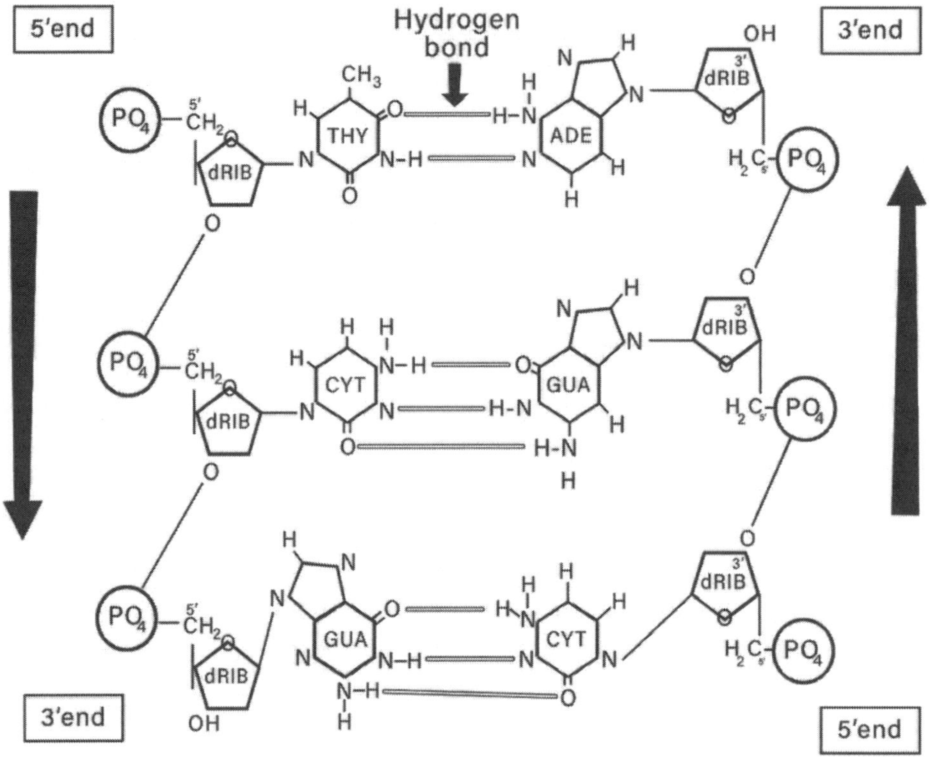

The dotted line between the bases in the middle are where these hydrogen bonds are located. Some bases have two hydrogen bonds between them, while others have three. You can bet that it takes more energy to break apart the bases connected with three hydrogen bonds than it does to break apart those with just two.

Proteins get their shape in part because of hydrogen bonding. The oxygen atoms and the hydrogen atoms on the amide parts of the protein will be attracted to one another through hydrogen bonding. In fact, hydrogen bonding will determine if a protein forms what's called an *alpha-helix shape* or a *beta-pleat shape.* In the same way, a protein made from two separate chains of amino acids will often be held together through intermolecular hydrogen bonding between the chains.

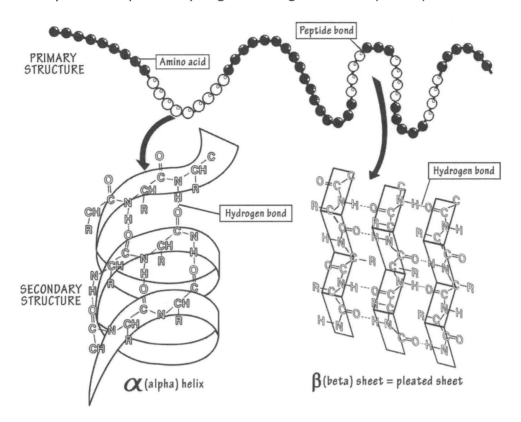

A related force to hydrogen bonding that you can get in molecules is called *van der Waals forces.* Like hydrogen bonding, these forces aren't truly bonds, but are more of an attractive force between two atoms. Van der Waals forces do not have to be attractive; they can be repulsive forces as well. A lot of reactions can be considered van der Waals forces but they each have different strengths.

The strength of the four different forces considered to be van der Waals forces include the following: Ionic bonding > hydrogen bonds > van der Waals (dipole-dipole) forces > van der Waals London dispersion forces.

Let's look at each of these:

We've already talked about ionic forces. These are the forces that exist between a cation and an anion. Two anions or two cations will repel each other because their charges are the same. On the other hand, an anion and cation are negatively and positively charged, respectively, so these two will attract one another.

According to Coulomb's law, the greater the charge on a nucleus, the stronger the attractive force between positively and negatively charged ions will be. In addition, as the distance between the two atoms increases, the attractive force will decrease.

We also talked about hydrogen bonding, which involves hydrogen plus an electronegative atom (in the same or a different molecule). Hydrogen has a medium electronegativity number of 2.2, so it really can't form a true ionic bond with anything, but it is enough to have some type of attractive force with other atoms. It will form hydrogen bonds in particular with atoms like fluorine, oxygen, and nitrogen because of the higher electronegativity of these atoms.

Van der Waals dipole-dipole interactions are those that do not involve hydrogen but have essentially the same phenomenon happening. It will involve any force between two atoms on different molecules (or that are far apart from one another on the same molecule) because of their differences in electronegativity. It's these differences that create a tiny bit of magnetism between the atoms so that they will either attract or repel each other. This is an example of how these dipole moments create an interaction between two atoms:

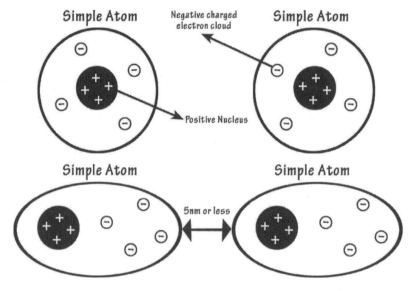

VAN DER WAAL'S FORCES DIAGRAM (VDW)
When two atoms come within 5 nanometers of each other, there will be a slight interaction
between them, thus causing polarity and a slight attraction

In the figure, the atoms get skewed like a magnet so that the positive nucleus of one atom gets attracted to the electrons of the other atom. Like two magnets, there is a tiny attractive force between them that tentatively keeps the atoms connected to one another in a very weak way.

Finally, the weakest forces are called *dispersion forces* or *London forces.* They're also types of van der Waals forces that are tricky to understand compared to some of the others. These are forces you'll see between instantaneous dipoles within or between molecules. It relies on the fact that atoms are never completely distributed around a nucleus. They are so random that, at any given point in time, it is almost a guarantee that some atoms will be unbalanced and will have more electrons in one spot in the atom. This creates a teeny tiny magnet between the positively charged nucleus and the negatively charged electrons.

It looks like this:

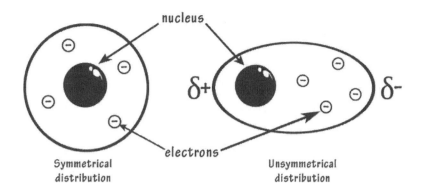

This dipole moment or "magnetism" is temporary but, if you put it on a large scale with zillions of atoms in a solution, you will get some attraction between the atoms going on. With an inert gas like argon, for example, it will do nothing as a gas, but as a very cold liquid, it holds together because of an imbalance between electrons in each of these atoms in the liquid solution.

The larger the atomic size of an atom, the greater the number of electrons it will have, and the more London dispersion forces you will see between the atoms. What this looks like on a large scale is that the atom as a liquid will take more energy or heat in order to boil it. This is a real thing. Argon (which is bigger by far than helium gas) boils at 87 degrees Kelvin (which is still very cold). Helium has a boiling point of just 1 degree Kelvin. The extra 86 degrees of temperature/energy necessary to boil argon is because it has more London forces going on than helium.

Fun Factoid: What do van der Waals forces have to do with geckos? A lot, as it turns out. Geckos stick nicely to walls and ceilings, but they do this without any real stickiness whatsoever. Instead, they have toe pads. On each toe pad, there are setae (very thin hairs). On each setae, there are even smaller spatulae. The spatulae have the ability to interact with the molecules of the wall or ceiling through van der Waals attractive forces, which make the gecko stick to the wall! They only get away with this because they don't weigh very much and, with so many millions of setae and spatulae, the force sticking them to the wall is strong enough.

Molecular Structures

Just like everything in nature, the lowest energy state is always preferred. This is true for atoms as we have determined, and it is also true for molecules. The whole reason molecules exist is because it is energetically favorable for them to stay together. The molecule's preferred shape is also completely based on which shape involves the least amount of energy. In some cases, the molecule will line up in a linear fashion, but this isn't true in others. Let's take a look.

Molecules can be linear. The bond angle for the molecule will be 180 degrees in order to keep all the charges that repel each other, including all of the electrons in the molecule, as far apart from one another as possible. One example of this is carbon dioxide:

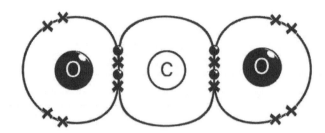

Another common shape is called the *trigonal planar shape*. These are flat and triangular molecules with bond angles that are exactly 120 degrees apart from one another. One example is sulfur trioxide, which looks like this:

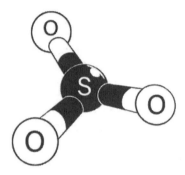

There are bent or angular molecules that look like they should be linear, but they aren't because the energy wouldn't be low enough in that shape. An example of an angular molecule is water, or H2O. The bond angle is 105 degrees. The hydrogen atoms are attached to two spots, while the lone electron pairs (two sets of them) take up the other two spots to make a four-pronged molecule that appears like it has two bent hydrogen atoms attached. It looks like this:

Water
Molecule

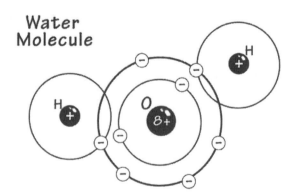

As we will talk about in a minute, the four outer electron pairs count in the overall structure. They do not like each other (except the two in the same orbital that have opposite spins, that is), so they also want to be far apart from the other pairs of electrons. The best way to do this is to create a bent molecule that takes these unpaired electrons into account.

The last common shape seen in biological molecules is called the *tetrahedral shape.* The term tetrahedral means "four faces," which translates into a molecule with four spots to fill. There are no unshared electrons, and the bond angle is 109.47 degrees. One example of this is methane, or CH4. It looks like this:

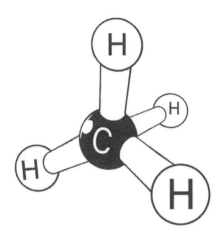

There are other possible shapes, such as trigonal pyramidal and octahedral, which aren't very common in biological systems. Ammonia, for example, is trigonal pyramidal because it has one pair of unmatched or unpaired electrons that has to be factored into the molecular shape.

The shape of any molecule is based on what's called the *VSEPR theory* or "valence shell electron pair repulson theory." It was proposed to explain molecular shape in 1939 and was later refined into the theory we use today. The concept is simple: All of the outer valence electron pairs count in the total molecular structure. These tend to repel each other, so the best structure is any 3-dimensional shape that keeps these repulsive forces to a minimum. Low repulsive forces equal a nice, low energy state for the molecule.

To Sum Things Up

Your study of atoms in chapter five prepared you for the study of molecules in this chapter. *Molecules* are created out of two or more atoms. A molecule can be as small as two hydrogen atoms or as large as a protein or piece of nucleic acid in your chromosomes. Molecules will form when it is energetically favorable for it to do so.

In order to form a molecule, you need to have *bonding*. Bonding happens when electrons are shared in some way between two atoms. When you see or hear that a chemical bond was formed, imagine an attraction of two nuclei to another one based on the sharing of their electrons. The electrons do not have to be shared equally (and in fact, equal sharing of electrons is usually the exception and not the rule).

The different types of bonds in a molecule vary in their strength and characteristics. *Covalent bonding* usually involves roughly equal sharing of electrons. There is a range, however, from nonpolar (pretty equal sharing) to polar (somewhat unequal sharing) covalent bonds. *Ionic bonding* involves extreme polarity, where there is a large difference in electronegativity between the two atoms involved.

There are weaker forces between atoms and molecules that aren't technically bonds but are more like attractive or repulsive forces. *Hydrogen bonding, van der Waals dipole-dipole forces,* and *London dispersion forces* are all weak forces that can exist. These forces are a lot like magnetism on a very, very small scale.

Molecules will take on the shape that is the most energetically favorable. The goal with any molecular shape is to have all of the electron pairs (whether they are involved in a bond or not) as far apart from one another in three-dimensional space as possible.

CHAPTER 7:
WATER, SOLUTIONS, AND MIXTURES

The next step in your understanding of how biological systems work is to learn about water and what molecules look like in a solution of water. Contrary to the belief that we are 99 percent water, the truth is that our bodies are between 50 and 60 percent water. That's still a lot, which makes it important to understand water as a solvent. Most of your body's biochemical reactions take place in water. In this section, we will also talk about some important features of watery solutions, such as molarity and acid-base chemistry in your body's watery (or *aqueous*) environment.

Water as a Unique Molecule

Water is an amazing and unique molecule with properties that make it the best possible solvent for the biochemistry of your body. As you know, it is a small molecule (just H2O). It is also polar, meaning it has a dipole moment that will help each molecule to attract to other molecules and to other polar molecules. Chemists call water the "universal solvent" because it dissolves so many things. It is also the solvent of living systems, being the most abundant molecular substance on the surface of the earth.

A lot of what makes the most sense in cell biology and in the biochemistry of living things happens because of the chemical properties of water. It is liquid at room temperature and normal pressures on the Earth's surface, but, as you already know, it can easily exist on our planet as a solid (ice) or as a gaseous substance (water vapor or steam). This wide range of possibilities and the features of water in these different phases make water the best possible solvent for life.

On Earth, most of the water you'll find is in the liquid state. As a liquid, water molecules float around, attracted to one another through hydrogen bonding. This image shows you how hydrogen bonding works in the different phases of water:

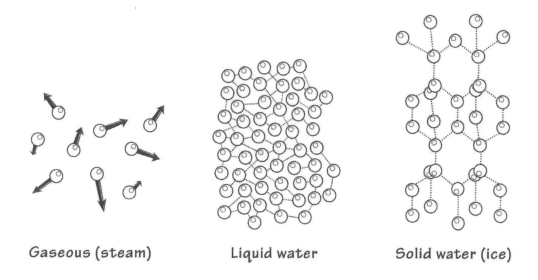

Gaseous (steam)   Liquid water   Solid water (ice)

Water as ice is also unique. Because of its hydrogen bonding, it forms a nice lattice shape that separates the water molecules out. This makes ice less dense than water, so it floats on the surface. This is a unique property of water that doesn't exist with all molecules in their liquid and solid forms.

Fun Factoid: Why would this idea of ice being less dense than liquid water be important? Imagine you are a fish in a pond just trying to get through the various seasons of summer and winter. As it got colder, ice would form on the pond. If ice happened to be denser than liquid water, it would sink, and you would get a solid block of pond water during the winter as the temperatures dropped. As a fish, you would find this to be very bad because you would basically die in a heap of frozen solid water at the bottom of the pond. As it is, ice floats and insulates the lower regions of the pond, so the fish have a fighting chance to make it through the winter deep beneath this icy blanket.

All molecules have a certain heat capacity. The specific heat capacity of a molecule is the amount of energy or heat (measured in *Joules*, which is an energy unit in physics and chemistry) you would have to apply to a substance to raise its temperature by one degree Celsius. Water has a very high specific heat capacity. This means it really resists any temperature changes; it sort of insulates itself from a change in its temperature as you apply heat to it. This is significant because most biological systems do not react well to large temperature fluctuations over time.

The high heat capacity of water is particularly important when it comes to *phase changes*. A phase change would be turning ice to liquid water or turning water to water vapor or steam, for example. The heat of vaporization is the amount of heat necessary to boil water; this is also high, largely because of the hydrogen bonding between the molecules.

In fact, it takes more energy to get water from 99.5 degrees Celsius to 100.5 degrees Celsius (the boiling point of water is between these numbers at 100 degrees Celsius) than it takes to increase the temperature of water the same one degree Celsius at any other point in its liquid phase (at room temperature, for example).

It's because of water that the Earth's temperature is a moderate as it is. The temperature extremes on earth would be much greater if it weren't for our oceans holding onto insulating energy to keep the temperature relatively stable. The same is true for the microcosm of your body and cells. The biochemical reactions in each cell depend on a stable temperature environment. Water itself helps create this important stability.

Water also resists being melted, so the amount of heat necessary to do this is greater than it takes to increase the temperature of ice or water while still in the same phase. This resistance to change from ice to water is why our glaciers are kept from melting to some degree and why ice is the best way to keep things cool for a long period of time.

Liquid water has a density of one gram per cubic centimeter. This is how researchers invented the idea of a gram in the first place. They used water as the main reference point. Water is less dense as ice, forming a hexagonal lattice shape. It is about nine percent less dense than water, so it floats nicely on a watery surface. There is a difference in the density of ice as it gets colder than its freezing point, which is 0 degrees Celsius.

Plain water and water with ions like table salt or sodium chloride do not behave the same. When you add salt to your water before boiling it, you actually raise the boiling temperature of the water but really not by any value you would notice. You would have to add nearly seven tablespoons of salt per liter of water in order to raise the boiling point of your salty water by one degree Celsius.

Salt added to water lowers the freezing point. The amount of salt in the ocean generally lowers its ability to freeze by about 1.9 degrees Celsius. It also lowers the density of water. This is really important in the world's oceans. Near the poles (where water is cold), water sinks, but the ice on top is nearly free of salt. The salt then collects in the watery parts of the cold ocean, so this water is saltier than the ocean water near the equator.

We will talk about the difference between solutions and mixtures in a minute but first, let's discuss the term *miscible.* Miscible means that a liquid substance will mix with other solvents or liquids. Water is miscible with a lot of other liquids, including ethanol/alcohol, glycerol, acetone, and acidic liquids. It is *immiscible* (not mixable) with oils or lipids. This is important to cell biology and to the biochemistry of your body. In general, polar solvents will be miscible with one another but polar and nonpolar liquids will not be miscible.

Water as a gas (water vapor) is miscible with air but it depends on the temperature of the air. As you cool the air, water will be unable to stay in its gaseous form and you will get condensation. What does condensation look like? Clouds and dew, basically. It is why you get dew on the grass in the morning and high humidity (high water vapor concentration) on warmer days.

While we talk about water vapor and steam as though they are the same, they really aren't. You can't see water vapor, but you can see steam, which involves tiny droplets of liquid water suspended in air. Steam you can see is called *wet steam*; this is what we think of when we see boiling or hot water as it turns into a gas. Higher temperatures will make all steam gaseous, turning it into water vapor, which is also called *superheated stream*. This is what makes the idea of steam versus water vapor so complicated.

Another interesting property of water is that it conducts electricity. Electricity is basically the movement of electrons from one place to another. (Technically, protons can conduct electricity, but this is rare and hard to do. It only happens in solids with positively charged hydrogen protons conducting the electricity.) So many properties of cells involve some kind of electricity, but the chemical kind rather than the wiring kind of electricity. Pure water without any salt ions is slightly electrical and can conduct electricity. But if you add any kind of salt to it, water as a salt solution is very electrical and conducts electricity extremely well.

Water is also very "sticky" or cohesive. This is because of its hydrogen bonding. It means that water molecules like to stick together. You can easily see how this works by looking at a drop of water on a surface; it bulges up as it tries not to spread out in any way. If you took a drop of ethanol or acetone instead, you wouldn't see the same droplet formation.

There are actually two related terms here. When we talk about *cohesion,* we mean that a substance sticks well to itself. The word *adhesion* means that something sticks well to other things. Water is both cohesive and adhesive. Its adhesiveness makes water droplets stick to glass. Throw some ethanol on a glass window and you won't see those droplets as much. Adhesion is in place inside cells; it's what allows water to be attracted to the cell membrane's polar surface. The hydrogen bonding in water also gives it a high surface tension, which helps insects walk on water.

Fun Factoid: It is because of the adhesion and cohesion of water that you can get water up to the tops of trees as high as a hundred meters off the ground. These forces create what's called capillary action so that water can be sucked upward in tiny plant vessels up through the plant's interior. If the vessels or tubes are too wide, the force of gravity or the weight of the water would overcome these forces. The tubes have to be narrow for this to work.

pH and Acid-Base Issues

You've probably heard about the *pH* of something and about acids and bases. These concepts are extremely important to biochemistry and to the function of all living things. We use the term "pH" to talk about watery or aqueous solutions. Theoretically, it could be applied to non-water solutions as well, but this is never done in practice.

The term pH is an easy way to describe the concentration of hydrogen atoms or ions in a solution. It is a convention we use to describe how acidic or alkaline a substance is. (You should know that the terms *basic* and *alkaline* are used to mean the same thing.) The pH scale goes from 0 to 14 and is essentially the negative log of the hydrogen ion concentration. A hydrogen ion is "acid" in chemical terms. It looks like this:

pH = -log[H+]

The term *log* (or *logarithmic*) is a fancy way of saying power. The log of 10^2 or 100 is 2. This is because it takes 2 multiples of 10, or 10 x 10, to get to 100. So, as it applies to pH, if the concentration of hydrogen ion is 10^{-14} moles per liter, the log of this would be 14. This is an extremely tiny quantity of hydrogen ions, so it wouldn't be acidic at all and would be basic or alkaline instead. If the hydrogen ion concentration is 10^{-1} moles per liter, the pH would be 1, which is very acidic.

This is what the pH scale looks like:

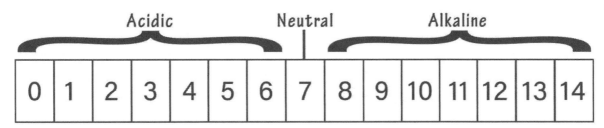

The term "neutral" in biological systems is a pH of 7. "Acidic" is anything with a pH less than 7 and "alkaline" or "basic" is anything with a pH greater than 7. Technically, it is possible to have a pH of less than 0 or greater than 14 but these numbers aren't really important for practical use.

On the other hand, the hydroxyl ion, or OH ion, is the classical base, just as the hydrogen ion is the classical acid. As you will see, it is a bit more complicated than that but it's still a good reference point for most biological systems.

Fun Factoid: Where did they come up with the term "pH" anyway? It was originally named by a Danish chemist (Søren Peder Lauritz Sørensen), who worked at the Carlsberg Laboratory in 1909. The term pH means the "power of hydrogen." This is where the H comes from. The origin of the "p," however, is up for grabs. Germans say it stands for the German word for power,which is "potenz." The French say it stands for the French word for power, which is "puissance." But it could be simpler than that. Sørensen labeled his acidic or positive electrode "p" and his negative or alkaline electrode "q." The "p" could simply refer to his electrode convention.

Now that you know what pH stands for, let's talk about what an acid is versus what you would call a base. It turns out there are three separate ways to define them. There are three theories on what acids and bases are — all of which are valid.

A chemist named Svante Arrhenius came up with the the Arrhenius theory on acids and bases. According to Arrhenius, an acid is anything that gives up hydrogen ions in water in order to lower the pH. This would include something like HCl or hydrochloric acid, which dissolves into hydrogen ions and chloride ions in water. Bases are anything that dissolves in water and gives off a hydroxyl ion. Sodium hydroxide, for example, divides into sodium ions and hydroxyl ions when it dissolves in water.

You need to know first whether or not water itself is an acid or a base. The answer is that it is both. Water as a molecule has the chemical formula of H_2O. When it breaks down, it divides into a hydrogen ion and a hydroxyl ion (sort of). In reality, there is no such thing as hydrogen ions floating free in water.

One single hydrogen ion is a proton, which by itself is only one hundred-millionth as big as even the smallest atom. The size of the atom does not really matter if its charge is powerful enough.

In water, hydrogen attracts easily to the water molecule itself, forming what's called a *hydronium ion* or H3O+ ion. This is an example of water acting as a base because it accepts hydrogen into itself. This image shows what this looks like:

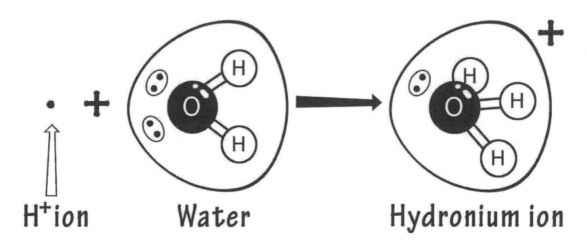

H^+ ion   Water   Hydronium ion

This is such as strong connection that there just aren't any actual free hydrogen ions anywhere in water.

Arrhenius had the right idea but not completely. The idea behind it is that an acid must give off hydrogen ions and a base must give off hydroxyl ions, but this just isn't true. Ammonia, for example, is a weak base. Its chemical formula is NH3, and in water, it gives off hydroxide or OH- ions but doesn't actually have this as part of its formula. There needs to be a better definition of acids and bases, then.

A few years later, two chemists named Johannes Nicolaus Brønsted and Thomas Martin Lowry came up with a new theory in 1923 but they didn't actually work together. Their theory is called the Brønsted-Lowry theory of acids and bases and is based on whether or not a compound will donate or accept a proton (remember that a proton is a hydrogen ion by itself). An acid donates protons and a base accepts protons.

This leads to the idea that something can be both an acid and a base, depending on the circumstances. A molecule that does this is called an *amphoteric molecule*. Ammonia is a base because it accepts a proton so that NH3 becomes NH4+. If you mix hydrochloric acid and ammonia, you get a reaction like this one:

HCl + NH3 → NH4+ + Cl-

In this case, chloride is referred to as the *conjugate base* of hydrochloric acid and NH4+ is referred to as the *conjugate acid* of NH3 (which is the base in this equation). All acids will have a conjugate base and all bases will have a conjugate acid to go along with it.

There are also strong acids and weak acids as well as strong bases and weak bases. The difference is how these *dissociate*, or separate, into parts in water. Strong acids and bases dissociate completely, while weak ones only dissociate incompletely in water. In biological systems, most acids and bases you'll see are weak. Only acids and bases like hydrochloric acid, sulfuric acid, and sodium hyroxide are strong acids and bases. Acetic acid or vinegar, on the other hand, is a weak acid. Ammonia is a weak base.

Another theory was also put forth in 1923. This time it came from Gilbert Lewis, the same chemist who came up with the Lewis dot structure of molecules. He used what he knew about chemical bonding to change things up a little bit. Instead of proton transfer, he based his theory on electron transfer. With this theory, an acid is anything that accepts a pair of electrons, while a base is anything that donates a pair of electrons. Acids are called *electrophiles* and bases are called *neutrophiles*. Acids have at least one spare or empty orbital, while bases have a set of unpaired electrons.

This opens up the concept of "acids" and "bases" widely. All cations, like copper and iron ions, are called *Lewis acids* because they accept electrons. In living systems, carbon dioxide (CO_2) is acidic because it accepts water to make carbonic acid. This happens in your red blood cells all the time and explains how some of the carbon dioxide in your blood gets carried through the bloodstream from the tissues to the lungs.

Molecules in a Watery Solution

We've talked about water being a universal solvent and, for biological systems, this is essentially the only solvent you will need to know about. A solvent does not have to be water, though. Any liquid that can dissolve another substance can be a solvent. A *solute* is generally a solid, but it doesn't have to be. Non-water solvents include organic substances, such as turpentine and benzene.

As we've discussed, polar substances dissolve in polar solvents, while nonpolar substances dissolve in nonpolar substances. A nonpolar substance like oil, for example, won't mix or dissolve in water. This is important in your cells. It explains why lipids and cell membranes (which are based on lipids) don't just melt into the watery environment of the cell. When you mix a solvent and a solute, you get a solution. Since the solute dissolves, the solution will be clear and not cloudy in any way.

What does a solution of salt in water look like? Many solutions in living things will be some kind of salt solution, like sodium chloride, or NaCl.

Water will interact with these ions to make what's called a *hydration shell* around each ion. Water sticks to each ion because of the polar nature of water. This is what it looks like:

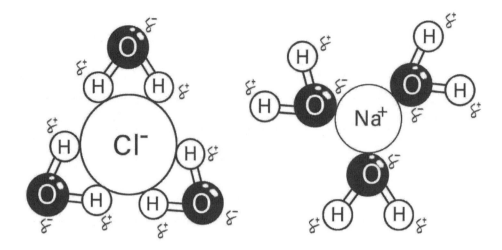

The ions are kept separate and dissolved in water by the water molecules surrounding each one. Ions in a solution are called *electrolytes* because they can conduct electricity. Free ions, like sodium, potassium, chloride, carbonates, and phosphates, are all common ions in biological systems.

You can measure the concentration of something in a solution in several ways. You can say that the sodium concentration in your blood is a certain number of milligrams of sodium or a certain number of milliequivalents of sodium per milliliter. What does this mean, anyway?

If you think about it, a milligram of potassium, for example, won't involve the same number of atoms as a milligram of potassium. Why? Because their atomic weights are different. The atomic weight of sodium is 22.98977 grams per mole and the atomic weight of potassium is 39.0983 per mole. This means that potassium weighs nearly twice as much as sodium per atom. The milligram-to-milligram comparison just doesn't make a lot of sense in biochemistry because of this.

Backing up a bit, let's find out what a *mole* is in the first place. One mole is a number used to describe the number of atoms or molecules there are in something. One mole is the same as 6.02×10^{23} atoms. If a molecule of something can't be divided up (as happens with salts), a mole is the same as the number of molecules. The atomic mass is basically the same thing as the molar mass, which would be how much a mole of a substance would weigh.

The atomic or molar mass will be close to the number of protons plus neutrons in an element, but it won't be an exact number. For example, neon has 10 neutrons and 10 protons nearly all the time, which would give it an atomic mass of 20. The listed atomic mass, however, is 20.179 grams per mole. This is because they take into account that for every common atom, there will be a few isotopes with an extra neutron or so. This throws off the atomic mass.

Fun Factoid: Who figured out that there were so many atoms per mole? This number, 6.02×10^{23}, is called Avogadro's number or Avogadro's constant, named after Amadeo Avogadro, who was the Italian lawyer and scientist in the early 1800s who first came up with the idea that this type of constant existed (although he never invented the number itself). Another scientist after him tried to do this. Josef

What this means is that one mole of sodium chloride will weigh something different than another mole of potassium permanganate, for example. Moles are easier to figure out in chemical equations because they are balanced with respect to the number of atoms but not the masses of the atoms (which don't mean much in chemistry equations anyway).

But, as you may have seen, things like the sodium content of your blood will be listed in milliequivalents per liter rather than moles. Equivalents account for the fact that sodium chloride, as an example, dissolves into two separate ions. They call this the *valence* of an ionic compound, so the valence of sodium chloride will be two. We use milliequivalents instead of equivalents because equivalents would be a big number. The equation looks like this:

$$Milliequivalents = \frac{Mass \; x \; Valence}{Molecular \; mass}$$

It accounts for the actual mass of something plus its valence number plus its molecular or atomic mass. You will see this in biological systems where the concentration of something is often listed in mEq/L, or milliequivalents per liter.

This leads us to the final concepts: *molarity* and *molality*. The *molality* of a solution is the number of moles of a substance per kilogram of the solvent. The *molarity* is the number of moles per liter. Because one liter of water weighs about a kilogram, these two concepts aren't very different from one another.

What Are Mixtures?

A *mixture* is anything that has two different substances in it. Solutions, suspensions, and colloids are all mixtures, even though they are different from one another. You know what a solution is already and that it involves a substance dissolving in another substance, called the *solvent*.

A suspension is when there are tiny particles of something in a solvent, but they aren't dissolved and can settle out. Sand in water is an example of a suspension. A *colloid* is similar, but the particles are tinier than with a suspension so, if they settle at all, it happens slowly. Milk is a colloid, also called a *colloidal suspension.*

Gases can mix together in a solution of other gases. A colloid of a gas and liquid would be a liquid aerosol spray, while bigger droplets in a spray would be a suspension. Dust in the air is a suspension of a solid in a gas. Oxygen gas can dissolve in a solution of water, but whipped cream and the head on your beer are colloids or suspensions. Emulsions of two liquids are seen with mayonnaise and vinaigrette. These are colloids or suspensions, depending on the particle size. Styrofoam is a colloidal mixture of a gas and a solid.

To Sum Things Up

Water is a unique molecule that makes it the perfect solvent for living things. Because of hydrogen bonding, it has a higher boiling point than you'd think and its solid form is less dense than its liquid form. There are unique properties of salts in water that affect its density and other properties. These are important to know because most living things are ionic or salt solutions rather than pure water.

Acids and bases are important concepts in both chemistry and cell biology because a lot of biochemical phenomena in cells are based on the properties of acids and bases or on reactions between these two. In this section, we used three different definitions of acids and bases that all make sense in certain circumstances.

Mixtures can be solutions, colloids, or suspensions. In living systems, your cells have a lot of things in a watery or aqueous solution. Now that you know how the concentration of substances in a solvent is measured, you will better understand what it means when you see something measured in milliequivalents per liter or what a mole of a substance really is.

CHAPTER 8:
WHICH ELEMENTS ARE IN CELLS?

There are many elements, but fewer than 100 of them can be found naturally on earth. The rest have been manufactured and often have a very short half-life. Of these elements, about 25 can be found inside living cells. The top four elements make up 97 percent of the body mass of living people. These are oxygen, carbon, nitrogen, and hydrogen. Rounding out the top six are sulfur and phosphorus, which are also common in life. This image shows you the percentage of the major elements in your body:

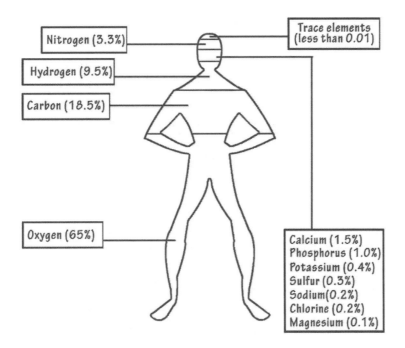

These few elements make up the thousands of different individual molecule types in each cell. We will talk about the large biomolecules that make up most of the molecule types in your cells. Carbon atoms are particularly common because they have four bonding sites, forming the backbone of many organic molecules. This is why scientists say that life is essentially *carbon-based.*

Elements that are important but aren't a major part of the larger biomolecules in life are chlorine, potassium, sodium, fluorine, iron, calcium, and magnesium. These tend to be important because they are charged particles or *ions.* Ions are extremely important in establishing charge differences across membranes, transporting molecules across membranes, and in providing cofactors for enzymatic reactions, among other functions.

Sodium and potassium, for example, are the main electrolytes used for the transmission of action potentials or *nerve impulses* in your nerves and are the ions used to help muscles contract (along with calcium). Calcium ions are also helpful for the clotting of blood. Fluorine and calcium make enamel, which gives teeth their strength. This is what enamel, which is mostly made of molecules of hydroxyapatite, are made of:

$$
\begin{array}{c}
Ca^{++} \\
^{++}Ca\text{--}O^- \quad O^- \quad O^- \quad O^-\text{--}Ca^{++} \\
O^-\text{--}P \qquad P\text{--}O^- \\
[^-OH] \qquad \qquad [^-OH] \\
\| \qquad \| \\
O \quad O \\
O^- \qquad \qquad O^- \\
^{++}Ca\,P\!=\!O\text{---}Ca^{++}\text{--}O\!=\!P\,Ca^{++} \\
O^- \qquad \qquad O^- \\
O^- \quad O \quad O \quad O^- \\
^{++}Ca\text{--}O^- \qquad P\text{--}O^-\text{--}Ca^{++} \\
O^- \qquad O^- \\
O^- \qquad \qquad O^- \\
^{++}Ca \qquad \qquad Ca^{++}
\end{array}
$$

Iron is another trace element in the body, making up only 0.004 percent of its total mass. Still, without iron, you cannot carry oxygen sufficiently through your bloodstream. Only a small amount of oxygen will dissolve directly in blood. Most is carried on the protein complex called *hemoglobin*, which contains iron necessary for it to hold onto oxygen. Another trace element is iodine. Without iodine, your thyroid gland will not function.

There are six additional elements that you should be familiar with, as these are the most abundant in living systems, especially when it comes to the making of many of the macromolecules (large molecules) in nature. The top six are: oxygen, hydrogen, carbon, nitrogen, phosphorus, and calcium. Let's study these elements to see what they do for your cells.

Oxygen

While life is said to be carbon-based, oxygen is the most common element inside your body. It makes up about 65 percent of the total body composition. In the Earth's crust, it is the most abundant element. However, it isn't the most abundant gas in our atmosphere. Oxygen only makes up about 21 percent of all of our air. (Nitrogen gas makes up the majority of our atmosphere.) As an atom, oxygen has two binding sites in most biological molecules.

Oxygen became such a common element on Earth largely because of our oceans. Oceans are made of water, which has oxygen in it as its heaviest element. Oxygen as a gas is vital to human life and the life of all aerobic organisms because it is crucial for producing a great deal of cellular energy. You take in food, break it down, and send the simple molecules to the cells. Inside the cells, these nutrients get broken down even further into energy for cellular functions. Oxygen is needed to do this effectively.

Fun Factoid: We didn't start out on this planet with much oxygen at all. Volcanoes in the life of early Earth created its atmosphere, which was probably mostly made of carbon dioxide plus some methane, water vapor, and ammonia. The water vapor is believed to have come from asteroids or comets, which have a lot of water in them. The Earth cooled, condensing water into the oceans and, somewhere in there, life began. As photosynthetic organisms arrived and gained a foothold on earth, the oxygen levels gradually rose.

We will talk a great deal about metabolism and cellular respiration in later chapters. *Cellular respiration* is the term used to describe using oxygen to metabolize glucose into sugar and also produce carbon dioxide, water, and energy. This requires oxygen and is the reason we can go for long jogs without being too winded or giving up altogether. You have energy for about 90 seconds without oxygen but, beyond that, cellular respiration kicks in and you are good for the long haul.

Fun Factoid: Your red blood cells contain the hemoglobin molecule, which carries oxygen to the cells. With all that oxygen, you'd think these cells would use aerobic respiration, but this is not the case. Red blood cells have no mitochondria, which is where cellular respiration takes place. Instead, they get so far in the energy-making process using biochemical pathways that don't generate a lot of energy but also don't need oxygen.

So, what happens if you are deprived of oxygen? If you are exercising, your muscle cells do have some energy reserve to carry them forth through activities like sprinting. Both sprinting and weight-lifting are called *anaerobic* processes because they don't use oxygen (the term *anaerobic* means "without oxygen"). The end product of this type of energy production is *lactic acid*. If your lactic acid builds up in the muscles, you'll feel soreness that takes a while to dissipate. If you go without oxygen for a long period of time (more than four to five minutes), you won't have the cellular energy to function and you suffer brain death and cardiorespiratory arrest.

Carbon

Carbon is most important as a backbone element. This is because it is present in all major biological/organic molecules and has four separate bonding sites per atom. This makes it a versatile molecule for biomolecules. You can find carbon as part of the backbone of every major biomolecule inside your cells, including proteins, carbohydrates, nucleic acids, and lipids/fats. As an atom, carbon is the fourth most common element in the universe.

Carbon cycles in two ways on Earth. There is a global carbon cycle, or geological carbon cycle, that takes millions of years to complete. On the other hand, the biological carbon cycle only takes a few days to up to 1,000 years to complete. Non-living carbon forms include carbon dioxide (present in very small quantities in the atmosphere), natural gas and petroleum products, dead organic material, coal, and rocks containing carbonate. In plants, carbon dioxide is used in photosynthesis to create organic molecules using the energy from sunlight.

The main molecule plants make is called *starch*, which is a polymer of many glucose sugar molecules in a linear structure. We (and most animals) eat glucose in this way and metabolize it, using oxygen to make carbon dioxide and water. The whole thing is brilliant because what plants give off, we need, and vice versa. The process is generally very balanced.

Fun Factoid: Global warming is a real thing, and it's a big mess for all of us. Because of pollution and even the gas given off by herd animals (farts, really), there are greenhouse gases released into the air. One of these greenhouse gases is carbon dioxide; it is called a greenhouse gas because it heats up the

Hydrogen

Hydrogen, as you know, is very simple. It is as simple as you can get in terms of elements, with one proton and one electron. Because of this simplicity, it has great bonding power but will only form a single bond with another atom. While it is small, it is the most abundant element in the universe. This is largely because hydrogen gas is part of what was given off in the Big Bang; it forms the clouds of dust and gases that make the stars. Stars will fuse hydrogen gas to provide the energy necessary to light and heat the planets. Planets like Jupiter and Saturn are called gas giants because they are made of hydrogen gas to a great degree.

Hydrogen in biological systems is found everywhere. It is in all major biomolecules and in water. It's a big part of the process that takes place inside mitochondria to provide energy. Hydrogen gas itself (H2 gas) is extremely abundant everywhere else in the universe but not as much as on Earth or in our atmosphere. It is extremely flammable, which makes it a great "star fuel." Some atomic bombs are hydrogen bombs, which use the fusion of hydrogen to create an explosion.

Fun Factoid: *If the fusion of hydrogen gas causes an explosion, why don't stars just explode? Because of the extreme mass of stars and the effects of high density and high pressure within stars (due to gravity), stars are unable to just explode—their gravity won't allow them to do that. Instead, it is an explosion in slow motion that allows stars to burn for millions and billions of years before running out of fuel.*

Nitrogen

Nitrogen is extremely abundant on Earth, making up 78 percent of our atmosphere. It is important for the growth and development of nearly all forms of plant life as part of the soil content. In living things, you'll find nitrogen in two of the four major biomolecules within cells — proteins and nucleic acids. In general, it has three binding sites in neutral molecules, such as you'll see with ammonia or NH3. Ammonium is another important ion on earth. This has the chemical formula of NH4+.

Of course, we wouldn't survive without nitrogen, so we need to eat foods that contain it. Fortunately, this is a big part of the molecular structure of proteins, so if you eat protein on a regular basis, you will get nitrogen to make your own proteins and nucleic acids. Foods like eggs, legumes, meat, and fish are all good protein sources. All life would end without it because you need nitrogen for your genetic material and for your enzymes and structural proteins.

Just as carbon has a cycle, so does nitrogen. It's called the *nitrogen cycle* and it's a big part of ecological systems. We, as biological organisms, take nitrogen in from our food to make amino acids and proteins. These will get broken down at some point by our bodies. The end metabolite, or byproduct, of protein degradation is *urea*. We excrete this urea in our urine. This is the structure of urea (note all of the nitrogen in it):

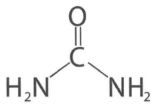

Plants also need a lot of nitrogen to survive. While they can take in carbon dioxide from the atmosphere, they cannot do the same thing with the nitrogen gas in the same atmosphere. For this reason, plants take nitrogen from the soil to make the chlorophyll pigment used for photosynthesis. They get this nitrogen from nitrates and ammonium ions from the soil. The nitrates and ammonia take their form because of bacteria in the environment that turn nitrogen gas into ammonia. The ammonia turns into nitrite ions, which is what plants use. The decomposition of plants and animals repopulate the nitrogen in the soil in the form of ammonia, where other bacteria turn it back into nitrogen gas. This *nitrogen cycle* looks like this:

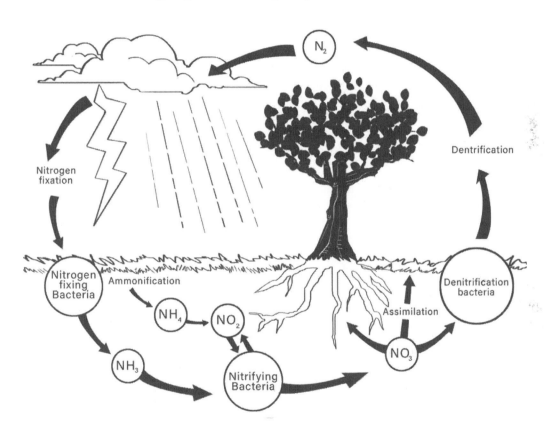

Sulfur

Sulfur is an important component of proteins. It's not present in all amino acids but it is found in both cysteine and methionine—just two of the 20 amino acids in living systems. These can be important for the structure of protein because the sulfur atom, which has two binding sites, will form a disulfide bridge between two sulfur-containing amino acids.

Sulfur is the seventh most commonly seen element on this planet. It is used by some bacteria as a major energy source for these organisms. In your body, you have about 140 grams of sulfur by mass. While your body can make all the cysteine you need, you cannot make methionine. This makes methionine an *essential amino acid* in your diet. Both methionine and cysteine need sulfite oxidase, which is a special enzyme necessary to break down these amino acids. The bridge affects the shape of the protein molecule. This is what it looks like:

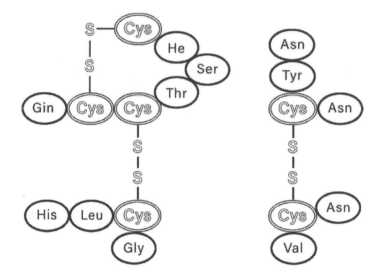

There are some common molecules that contain sulfur but are not amino acids or proteins. Two of the most common molecules that have sulfur are *coenzyme A* and *alpha-lipoic acid*. Coenzyme A is very important to your metabolism and alpha-lipoic acid is both an antioxidant and an important molecule for metabolism in several ways. The vitamins biotin and thiamine also contain some sulfur atoms as part of their structure.

Phosphorus

Phosphorus is crucial to life on Earth. The cell membranes of every cell contain *phospholipids*, which are long carbon molecules that have a phosphate molecule on the end. This is important because phosphate is polar, so it can form the outside face of the membrane on both the inside and outside of the cell. In the cell's interior, there will be a lipid tail that hides inside the membrane, which is double layered. We will talk about how membranes are structured in a later chapter. This is what a phospholipid looks like:

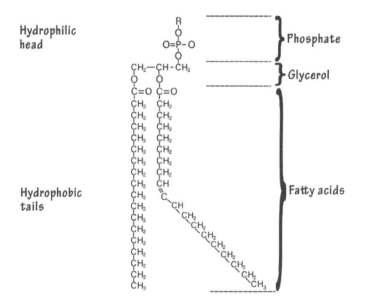

Notice in the image that the phosphate part is *hydrophilic* or water-loving, while the lipid/fatty acid tail is *hydrophobic* or water-hating. This is why you'll see all of the phosphate heads facing into and out of the cell membrane so it can interact with water and so the lipid tail doesn't have to do this.

Phosphorus is also crucial to the formation of the backbone of nucleic acids like DNA and RNA. We will talk more about their structure in a minute but for now you should know that the different subcomponents of DNA and RNA are connected through a phosphodiester bridge that connects each nucleotide to one another. This is a phosphodiester bond in nucleic acids:

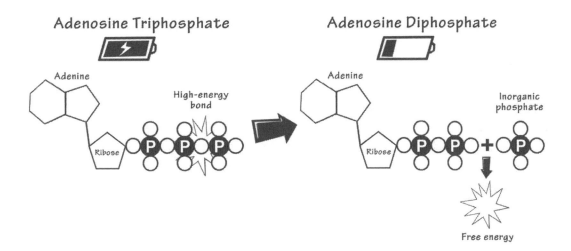

ATP, or *adenosine triphosphate*, is the major currency of energy in the cells. This has three phosphate molecules on it. When it breaks up to create energy, it does this by giving off one of its phosphates. When this happens, the process gives off energy used to drive biochemical reactions. This is what it looks like:

Phosphate can be found in bones as well. About 70 percent of your bones is a calcium salt called *calcium phosphate.* These salts are bonded together by a protein called collagen in your bone tissues.

To Sum Things Up

There are 25 known elements that make up living systems but the vast majority of these involve just six elements: *carbon, oxygen, hydrogen, nitrogen, phosphate,* and *sulfur.* Life on Earth is considered to be carbon-based, but oxygen is the most abundant element on the planet. In the universe, however, the major element is *gaseous hydrogen.* There are other elements not typically found in the major biomolecules that are still important ions. These include *calcium, magnesium, potassium, chlorine, sodium,* and *iron,* among many others.

CHAPTER 9:
MACROMOLECULES ARE THE
"BIG" MOLECULES IN LIVING THINGS

In this section, we will talk about the "big four," which are the four major biomolecules or macromolecules found in the cells of all living things. These include *carbohydrates/sugars, proteins, fats/lipids,* and *nucleic acids.* You already know a lot about the atoms involved in these molecules and how atoms themselves can form these huge molecules. There are key similarities and differences between these unique macromolecules.

Hopefully, by understanding the molecules themselves, you will know what they can do and will be better able to figure out how all of these molecules interact to form the machinery and biochemical reactions that make your cells work on a continual basis.

Carbohydrates

All carbohydrates are sugars, although there is some distinction between *simple sugars* and *complex sugars*, or complex carbohydrates. All sugars have only carbon, hydrogen, and oxygen molecules in them. Most have a 2:1 ratio of hydrogen to oxygen, and the most common sugars like glucose have a ratio of carbon to hydrogen to oxygen of 1:2:1. The most common sugar in your cells is *glucose*, which has a chemical formula of $C_6H_{12}O_6$.

When we say "sugar," you probably think of table sugar, which is a type of sugar called a *disaccharide*. There are four terms commonly used to describe carbohydrates:

- **Simple sugar** — You can call this a *simple carbohydrate* or a *monomer sugar. Monomer* means "single unit." They are also called *monosaccharides.* You'll see that some of the other macromolecules have monomers. When you put monomers together, you get *polymers.* Some common monomer sugars, or simple sugars, in living systems are *glucose, fructose,* and *galactose.* In nucleic acids, there are two simple sugars — *ribose* and *deoxyribose.*
- **Disaccharides** — These are common sugars that are made only of two simple sugars. Table sugar, or sucrose, is a combination of glucose and fructose, while lactose, or milk sugar, is a combination of galactose and glucose. Sucrose is also called cane sugar or beet sugar. Maltose or molasses sugar is a combination of two glucose molecules. As an example, take a look at what sucrose looks like:

- **Oligosaccharides** — The term *oligo* means "few," so an oligosaccharide is a polymer carbohydrate made from a few sugar monomers. An oligosaccharide by definition has three to six monomer sugars or monosaccharides in it.
- **Polysaccharide** — The term *poly* means "many," so a polysaccharide is a long chain of monosaccharides. There are many of these in nature. In animal cells, the main polysaccharide is called *glycogen* (made from glucose only). In plants, the main polysaccharide is *cellulose*, although some plants build up starch granules, which are also *polysaccharides*. Both amylose and *amylopectin* are types of starch in plants. Insects and mushrooms have *chitin* as their main polysaccharide. All except chitin are made from some type of glucose in a chain (differing only in how the glucose molecules are connected). Chitin is made from a chain of N-acetylglucosamine molecules, which are derivatives of glucose.

Carbohydrates are extremely important to our metabolism and to the metabolism of our cells. Even when you eat proteins and fats, these will not be the preferred energy source for your cells. Your brain only uses glucose as energy, so you need to have some of this in your diet or the mechanisms in place to turn the other nutrients you eat into glucose in order to fuel your brain cells.

Sugars are also part of structural and functional molecules in your cells. The outside of each cell has numerous glycoproteins on the cell surface. Glycoproteins have both sugar and protein together in one molecule; these are used for cell signaling and to help tag a cell for identification.

Ribose is a 5-carbon sugar (instead of the 6-carbon sugars like glucose and fructose). It is used as part of the nucleic acid RNA and to make ATP and other energy molecules, such as FAD and NAD. A related 5-carbon sugar is *deoxyribose*, which has a major role to play in making the nucleic acid DNA. There are many more sugar-based molecules used in blood clotting, fertilization, and immune function.

There are some things you might not think of as being a sugar or carbohydrate because they do not fit into the same structural pattern of typical sugars. Two of these are *sorbitol* and *mannitol*. Sorbitol is a common sweetener found in some fruits or synthesized from potato starch (the glucose molecules in the starch). It is not an artificial sweetener but is still sweet. It is called a "polyol" because it has multiple OH side chains, which are also called *alcohol side chains*. This is what sorbitol looks like:

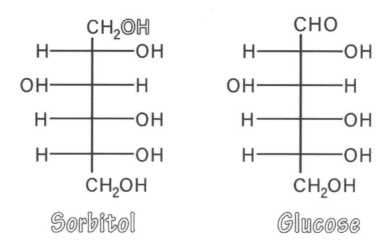

Sugars in nature are mostly called *D-sugars*. You will often hear about a molecule being a D-molecule or an L-molecule (and sugars are no exception). The letter "D" stands for *dextro* (which means "right") and the letter "L" stands for *levo* (which means "left"). In sugars, a left-handed sugar has a slightly different shape than a right-handed sugar. They look like this:

The convention in sugars is to base its handedness on the bottom hydroxyl or alcohol side chain. If it faces right, it is a *D-sugar*; if it faces left, it is an *L-sugar*. Two molecules with the same formula but different mirror-image shapes are called *enantiomers* of one another. This might not look that important, but it is. Enzymes and any other process in the cell that require two molecules to fit together depend on the handedness of the molecule. If a molecule is of the wrong handedness, it might not get recognized by enzymes and wouldn't get metabolized in the same way as the correct-handed sugar.

When we write the formula of any sugar, we can write it as we have already in a linear form but it can also be written in a ring form. This is because it can exist in nature in either of these forms and often switches back and forth between forms. In substances like sucrose, they are mostly in a ring form. Take a look at the sucrose image earlier in this chapter.

Notice that, on the left is glucose, which forms a hexagonal shape; on the right is fructose, which forms a pentagonal shape. Both have six carbon atoms per molecule, but in fructose, there is an extra carbon sugar that isn't included in the ring shape. You can call the glucose rings *pyranose rings* or you can call them *hexose sugars* because of the six carbon atoms in them. The fructose ring is called a *furanose* ring because it forms a 5-carbon ring.

When you eat any carbohydrate, it will provide you with about four kilocalories per gram of energy. The same is true of any protein you eat. Lipids or fats are more energy-dense, so they will provide you with about nine kilocalories per gram you eat. Carbs can be found in breads, fruits, vegetables, legumes, rice, and milk. About 45 to 65 percent of the energy you eat every day (remember that in nutrition, energy equals calories) should be some type of carbohydrate.

You can eat all the cellulose you want but, because your GI tract enzymes cannot break it down, it doesn't get counted as calories you can absorb. Instead, it becomes insoluble fiber that exits your bowels without getting absorbed as food. Other carbs, like sucrose or table sugar, are easily broken down and absorbed into the bloodstream. This means that you can eat more pounds of vegetables without having those calories count than you can eat of any refined sugar product, such as donuts, cakes, or bread. Now you see why they say to eat a lot of vegetables if you want to lose weight.

There is a lot of hype about low-carb diets these days, but are they healthy for you? Not according to medical science, anyway. You will miss out on the dietary fiber you get from whole fruits, vegetables, and beans, and there could be an increased risk for diseases like cancer and osteoporosis. Weight loss is about calories and not about carbs. The idea that carbs make you fat just isn't supported by the scientific research.

Proteins

Proteins are another type of macromolecule, or biomolecule, in your cells. Proteins are also called *polypeptides*. The monomer unit of a polypeptide is an amino acid. There are 20 different amino acids in living things that go into proteins but there are other possible amino acids that do not become a part of any protein molecule. The known elements that make up proteins are *carbon, oxygen, hydrogen, nitrogen,* and *sulfur.*

Proteins have so many different roles in the cell. Structural proteins make things, such as molecules in the cell membrane and numerous intracellular structures. The majority of the rest of the proteins in the cell are enzymes. An *enzyme* is any protein molecule that will perform work to help move a biochemical reaction along. You might think you are studying "cell biology" but really you are studying the biochemistry of your cells. You cannot make DNA, other proteins, or complex carbohydrates without enzymes. You also cannot break down molecules as part of metabolism without enzymes.

There are also proteins that perform miscellaneous functions. Small proteins act as *neurotransmitters* in the nervous system. There are proteins like albumin that bind to things in the bloodstream and there are proteins like hemoglobin that will carry oxygen in the red blood cells. Immunoglobulins are proteins that are important in the immune system and are called *antibodies.* Gluten is a protein you eat that is found in wheat and other grains.

As is true for carbohydrates, there are monomers and polymers involved here. The monomer of proteins is called an *amino acid.* Short chains of amino acids are called *oligopeptides*, while long chains of amino acids are called *polypeptides*, or just *proteins.*

Amino acids have different structures but there are similarities that make them amino acids. This is the basic structure of an amino acid:

As you can see, they have an *amine* group, or NH2 group, on one side and a *carboxyl*, or COOH, group on the other side. Next to the carboxyl group is a carbon atom called the *alpha carbon*. This carbon atom is important because it always has a side chain on it, called the *R side chain*. The side chain determines the name of the amino acid.

Side chains on the amino acid can be very simple. For example, the simplest amino acid is glycine because its R chain is just a hydrogen molecule. There are other side chains that are much more complex. These are the basic amino acid classifications:

Non-Polar	Carboxyl	Amine	Aromatic	Hydroxyl	Other
Alanine	Aspartic Acid	Arginine	Phenylalanine	Serine	Asparagine
Glycine	Glutamic Acid	Histidine	Tryptophan	Threonine	Cysteine
Isoleucine		Lysine	Tyrosine	Tyrosine	Glutamine
Leucine					Selenocysteine
Methionine					Pyrrolysine
Proline					
Valine					

Nonpolar amino acids will not like to be near water, so these tend to hide inside the interior of proteins or prefer a lipid-like situation. There are carboxyl amino acids that have a COOH in their side chain (and will often be acidic). *Amine amino acids* have a side chain with an NH2 group on it. *Aromatic amino acids* have what's called an *aromatic ring* as part of the side chain. These tend to be nonpolar. There are also those that have an OH, or hydroxyl, group and some miscellaneous amino acids. Both methionine and cysteine have sulfur atoms on the side chain.

When two amino acids, or *peptides*, combine to start the formation of a protein, they form a chemical bond between the COOH (carboxyl) group and the NH2 (amine) group. This bond is called a *peptide bond.* Many amino acids strung together with peptide bonds are called *polypeptides*. This is what a peptide bond looks like:

A string of a certain combination of amino acids is called the *protein backbone.* This is also known as the primary structure of the protein. The secondary structure happens based on the hydrogen bonding pattern between different amino acids in the chain. Depending on how these are bonded, you might get specific shapes, such as an alpha-helix shape or a beta-pleated shape. The tertiary structure of a protein is how the protein fits together in its final form because of interactions between the side chains. The disulfide bond between two cysteine amino acids will form a certain permanent shape to that segment of protein. The quaternary structure is the formation of a super-protein made by more than one polypeptide chain connected to each other. This is what each of these looks like:

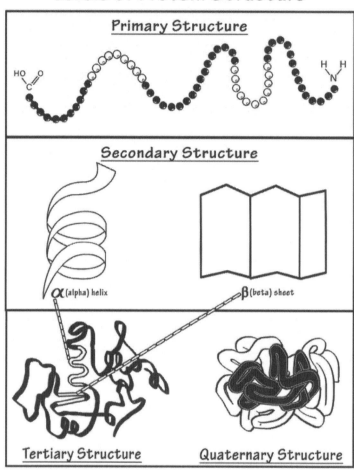

The process of making proteins is complex and starts with the DNA message for the protein, usually held in a single gene. This message gets transcribed and passed through a different molecule to the ribosomes of the cell, where they translate the nucleic acid message into a polypeptide chain using what's called the *genetic code.* Some of these proteins get modified even further to form the size and shape needed to do their specific functions.

When you eat foods with protein in them (like meat, fish, eggs, dairy, nuts, beans, and legumes), your body's GI tract breaks them down into *amino acids*. It's these amino acid building blocks that get absorbed into your system and sent to your cells to make brand new proteins. Some amino acids can also be synthesized from scratch while others can't. In humans, there are nine amino acids that are deemed "essential," which means they are an essential part of your diet. The amino acids essential for humans aren't the same amino acids essential for other organisms.

Lipids

Lipids are the third type of macromolecule, or biomolecule, we will talk about. These are the major nonpolar substances in your cells. In chemistry, you will find these able to dissolve in hydrocarbons like benzene or gasoline but never in water. There are a lot of different lipid types in your body and a few that aren't in your body at all. These are the different lipid types:

Fatty acids — When you eat fatty foods, you are mostly eating fatty acids. Your body also stores its fats in this form. A fatty acid is basically a long hydrocarbon chain plus a COOH (carboxyl) group on the end. The basic structure is this:

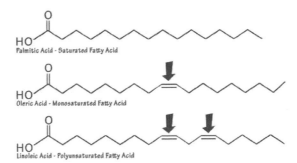

Essential features of a fatty acid

A *hydrocarbon* is essentially a long chain of varying lengths that just has carbon and hydrogen atoms on it. You've heard of saturated and unsaturated fats, right? These refer strictly to the structure of the hydrocarbon chain on a fatty acid molecule. A *saturated fatty acid* is one that is saturated with hydrogen ions in every possible spot. An *unsaturated fatty acid* does not have the total number of possible hydrogen ions. Instead, there are double bonds between at least one pair of carbon atoms. There are *polyunsaturated fats* (with many double bonds) and *monounsaturated fats* (with one double bond). They look like this:

Notice how the double bonds look dome-shaped? These are called *cis-fatty acids* because the two hydrogen atoms (or two long chains on either side) are on the same side of the double bond. Trans-fats are the opposite. They look like this:

cis-fatty acid

trans-fatty acid

Trans-fats were invented by food chemists because they resist spoiling when put into food. The problem is that they will raise your bad cholesterol levels and promote heart disease. Don't eat them!

Fun Factoid: *In 1901, a German chemist named Wilhelm Normann invented a way of hydrogenating liquid oil by mixing it with hydrogen gas. This process made a semi-solid fat that was mostly trans-fat. He got a patent for this and it was soon added to foods in order to keep them from spoiling. While doctors had linked saturated fats to heart disease in the late 1950s, they didn't realize that trans-fats were worse for you until the 1990s. Since then, as many food manufacturers as possible have been phasing out the use of trans fats. Trans fats aren't banned but food manufacturers must say how much trans fats are in a particular food you eat.*

You've heard about *omega-3* and *omega-6 fatty acids*, too, right? These are unsaturated fats that have their double bonds in certain places. The omega carbon on the fatty acid is the last one in line. An omega-3 fatty acid has its double bond between the third and fourth carbon atom from the omega carbon. Omega-3 fatty acids have many health benefits when we eat them. Both *alpha-linolenic acid* (an omega-3 fatty acid) and *linoleic acid* (an omega-6 fatty acid) are essential for the diet.

- **Waxes** — These are a diverse group of lipids that are solid and malleable at room temperature. They are named for their appearance more than they are for their chemical structure, which can involve a lot of different structures and side chains. Examples of waxes in nature that you might see include beeswax used to make honeycombs. The oil in the head of a sperm whale is a type of wax called spermaceti. Plants make waxes all the time in order to keep them water from evaporating.

- **Sterols** — Sterols are important to your cells and are basically fatty acids that have folded themselves into a specific series of intertwined rings. Cholesterol is one of the main sterols in your body, often found floating in your cells. The structure looks like this:

You can find cholesterol and other sterols in bile acids. Yes, you can eat cholesterol but most cholesterol is made by your liver (because you really need it). Cholesterol is not the dietary enemy you'd think it should be, but you can certainly take in too much of it. Only animals make cholesterol; its role is to keep the cell membrane from becoming too stiff.

Cholesterol is the root molecule for many different hormones in the body. *Estrogen, testosterone, progesterone, glucocorticoids,* and *mineralocorticoids* in your body are synthesized from cholesterol. Vitamin D is a type of sterol called a *secosteroids*. When you see the term *steroid* or *steroid hormone*, you should know they mean it looks somewhat like cholesterol and that it was probably derived from it.

- **Lipid-like Vitamins** — There are four fat-soluble vitamins you need in your diet. These are *vitamins A, D, E,* and *K.* They have different structures but are put together because they are soluble in fat. You need to eat some fat in your diet to get these vitamins in as well, and if you can't absorb fats for some reason, you may become deficient. An example of a fat-soluble vitamin is vitamin A, which looks like this:

- **Triglycerides** — These are basically a type of *fatty acid*. In fact, a triglyceride is three fatty acids connected to one another by an alcohol molecule called glycerol. You can have *monoglycerides* and *diglycerides,* too (which have either one or two fatty acids attached to the glycerol molecule, but these aren't as common). The fatty acid side chains do not have to be identical to have it be called a triglyceride. A triglyceride looks like this:

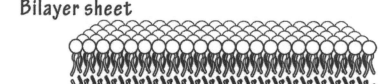

| glycerol | 3 fatty acids | triglyceride (triester of glycerol) |

The main function of triglycerides is to store fatty acids so that they can be broken down and used as fuel for the body's metabolic processes.

- **Phospholipids** — *Phospholipids* are basically fatty acids with a phosphate molecule at the end instead of a carboxyl group. The phosphate group is highly polar, and the fatty acid tail is highly nonpolar. These molecules make up the majority of what's in the membranes of all cells. The fact that there is a polar end and a nonpolar end means that they must line up in what's called a *lipid bilayer* so that only the phosphate heads are exposed. In a living cell membrane, this phospholipid bilayer looks like this:

Bilayer sheet

Lipids are great for fuel storage since they are so *calorie dense*. Because you need a long-term storage molecule, you have fat cells that will collect triglycerides until such time as you need them for energy. Glucose is stored in the liver as glycogen (another form of energy storage molecule) but your triglycerides are in it for the long haul and will help you have energy even if you haven't recently eaten.

All membranes, both outside the cell (the *cell membrane*) and inside the cell (*organelle membranes*) are made of lipids. There are different ways to expose these lipids to watery environments, such as using phosphate heads or parts of the lipid attached to carbohydrates, which are also *hydrophilic*. They mix well with water near the external surface of the cell membrane. Cholesterol, as you will see, also floats in the cell membrane structure. Fat-soluble vitamins are used in many cell functions.

Fatty acids get metabolized similar to how glucose gets metabolized. When glucose is exhausted, fatty acids are degraded through a process called *beta oxidation*. This breaks down the fatty acids into small pieces and feeds them into the same metabolic pathways that glucose goes through. This is part of how triglycerides/fatty acids get used to make ATP energy when the cell needs a long-term fuel source.

Nucleic Acids

Nucleic acids are the fourth and final type of large macromolecule in your cells. You know these types of molecules as the genetic material inside all of the cells in living things. The monomer involved in making nucleic acids is the *nucleotide*, which is made from one kind of 5-carbon sugar (ribose or deoxyribose), a nitrogenous base, and a phosphate group. You may have heard of a *nucleoside* as well. These are similar to nucleotides but do not have the phosphate group on them. This figure shows what these are:

Nucleosides vs. Nucleotides

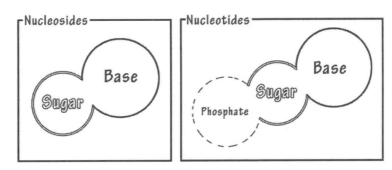

A nucleic acid is a polymer of nucleotides attached to one another through *phosphodiester bonds*. The term *nucleic acid* is used interchangeably for DNA (deoxyribonucleic acid) and RNA (ribonucleic acid), even though they are different. RNA uses a ribose sugar while DNA uses a deoxyribose sugar (which only has a hydrogen ion on the second carbon atom in the ring instead of a hydroxyl group).

There are five different nitrogenous bases in nucleic acids; usually, these are named according to the first letter of their name. These are *adenine, guanine, cytosine, uracil*, and *thymine.* DNA never has uracil and RNA never has thymine.

The other main differences between DNA and RNA are that DNA is mostly double-stranded and RNA is mostly single-stranded. DNA looks like a ladder with the sides made from the sugar and phosphate parts and the rungs made from two nitrogenous bases per rung. There is a specific way the bases bind to each other with hydrogen bonds. Adenine and thymine have one bond while guanine and cytosine have another.

This is what the DNA double helix looks like:

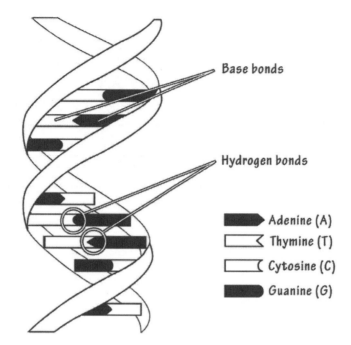

When looking at the structure and the way they line up, the two strands or sides of the ladder are called *antiparallel strands*; this is because one goes in one direction, called the *3' to 5' direction* and the other goes in the opposite *5' to 3' direction*. The 3' end is where the hydroxyl group on the deoxyribose sugar sits, while the 5' end on the same molecule has the phosphate group. This is what it looks like:

It turns out that when DNA lines up this way, it is more stable as a molecule.

RNA is often *single-stranded*, but it can easily fold on itself in what are called *hairpin turns*. At these turns, the polymer will do the same kind of bonding but with other molecules on the same strand, creating unique shapes based on the location of each nucleotides.

There are several kinds of RNA. Each has its own job. *Messenger RNA* takes a message from DNA inside the nucleus and passes it outside the nucleus to be translated into a protein chain. Ribosomal RNA is used to make the ribosomes that do the job of making proteins. *Transfer RNA* carries a single amino acid per molecule and helps to add to the polypeptide chain. There are small types of RNA, such as *small nuclear RNA* or *snRNA* that we will talk about soon. From this picture of transfer RNA, you can see how unique shapes can be made from a single strand of RNA:

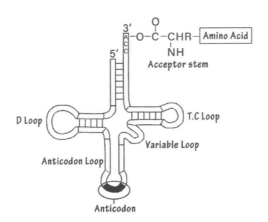

Nucleic acids are ingenious because they create long chains of different nucleotides that are part of the genetic message inside each cell. You can see where the different nucleotides tell a story in a kind of simplified alphabet that will be able to code for the thousands of different proteins your cell needs to make. These nucleic acids are basically instructions that are passed from one cell to the next generation of cells in order to preserve the types of proteins being made.

DNA in eukaryotic cells like our own are not usually naked pieces of DNA. They are linear strands of different chromosomes that are tied up with proteins to make DNA more condensed and easier to pack. These proteins are called *histone proteins* and have DNA wrapped around them in what's called a *nucleosome*. These are lined up along the chromosome like beads on a string. Their other function is to keep the wound-up segments of DNA from being transcribed into an RNA message. It is one way that we regulate which genes get turned on and which are suppressed.

Here's what these nucleosomes look like:

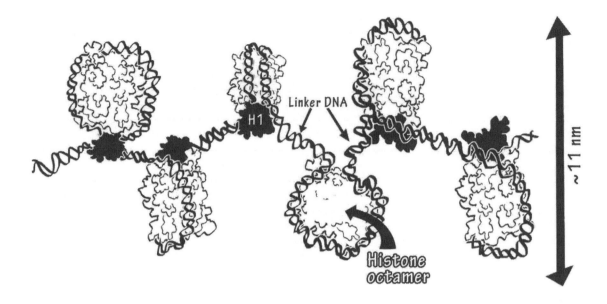

To Sum Things Up

In this section, we opened the door to a real discussion of cell biology by talking about the major molecules you'll see inside cells. Each of these is unique; most involve some kind of *monomer*, or "single unit," that is linked together with other monomers to make a *polymer*.

When we think of macronutrients in your diet, these are divided into *proteins, lipids,* and *carbohydrates.* Nucleic acids are macromolecules but aren't actually considered nutrients, even though you do eat them. Unlike the others, nucleic acids aren't consumed in great numbers and are made from scratch in your cells. You should know that carbohydrates and proteins have the same energy density, providing you with four kilocalories (one food calorie) per gram of substance. Lipids have more energy at nine kilocalories per gram.

SECTION THREE: ENERGY IN THE CELL

You should now understand a great deal now about how atoms become molecules and how molecules in the cell can be quite large in order to carry out cellular processes and establish the structure of the cell itself. All of this takes *energy,* which must come from some type of fuel. It's like your car. You use gasoline as fuel, but you need the rest of the engine in order to convert this fuel into useable energy.

In this section, you'll learn how the *thermodynamics* also applies to living systems. Thermodynamics is all about energy; there are plenty of energy systems in all cells. Energy can be used to build things or to break them down.

Let's first look at thermodynamics and how it applies to living systems. Then, we'll cover the energy molecules inside the cell and discuss why these are called *energy molecules.* Finally, we will talk about what cell metabolism looks like and how enzymes (mostly proteins) participate in metabolic processes.

CHAPTER 10:
THERMODYNAMICS IN LIVING THINGS

Strictly speaking, thermodynamics is the study of how temperature, heat, work, and energy are related. In reality, you can think of thermodynamics as being as close to the "theory of everything" as you can get. No matter what you are talking about — physics, chemistry, living systems, and even psychological systems — thermodynamics will apply.

While you can get bogged down in a lot of thermodynamics, it is probably best to think about it in terms of energy science, mostly because a lot of it has to do with how energy works in all kinds of systems. You can think about the entire universe as being a giant ball of energy. According to thermodynamic theory, the energy in the universe will always stay the same. It can never go up or down in any way, but it can change forms.

On a much smaller scale, imagine a piece of paper on a table just sitting there. It doesn't look like it has much energy in it, and it doesn't move. If you add a lighted match to it, it will burn and will become a pile of ash. Is it the same piece of paper now? Can you make a piece of paper out of the ashes again? What happened to its energy? People who understand thermodynamics will be able to answer these questions.

First, let's look at the laws of thermodynamics. They are used in physics, biology, and chemistry in order to understand how these different systems work. The four laws are:

- **The Zeroth Law of Thermodynamics** — This one is easy. It says that if you have three separate systems and two of these are of the same thermal state (temperature, basically), the third system must also be in the same state. It's used to explain how you can use a thermometer to measure two systems and, if you get the same value for each, the two systems must be thermally the same.

  Fun Factoid: Why call this the zeroth law? Well, historically, the first three laws of thermodynamics were developed in the early 1700s, but then in the early 1900s, a scientist named Ralph Fowler came up with a law that seemed to precede the others. Because of this, he decided not to call it the fourth law of thermodynamics but instead to call it the zeroth law.

- **The First Law of Thermodynamics** — This is also known as the "law of conservation of energy." It states that, in any energy system, no energy can be added or subtracted from the system itself. You can add energy from outside a system or lose energy from a system, but within a given system, the energy may change but will not be lost or gained. If you use a piston to do work on an engine, the energy in the system goes to do the work but the whole amount of possible work will not be perfect. There will be some energy given off as heat from friction. But, if you think of the airspace around the engine, that energy just turns into heat and still remains as part of the total system.

- **The Second Law of Thermodynamics** — This law talks about *entropy,* which is a measure of disorder in a system. It means that energy systems will always go from a state of low entropy to a higher (more disordered) state. Think of the piece of paper burning. The pile of ashes has a much higher entropy state than the piece of paper because it is more disordered. You also cannot have a cold area expand into being a hotter region. Think of heat as being energy and cold as being a lack of energy, or *anti-energy*. Energy can expand or become more disordered from hot to cold but not the other way around.

- **The Third Law of Thermodynamics** — Entropy drops as the temperature drops. All elements and molecules will eventually become solid as the temperature decreases so that the molecules will not move around. Essentially, even the electrons and nuclei will not move near absolute zero, which is -273 degrees Celsius. A substance (now solidified) will be at its lowest entropy state at that point. Again, without any heat, entropy or disorder will decline and the system will be less disordered.

You will see how energy is put into a biochemical system inside a cell but that the natural order of things in all systems is to settle into the lowest possible energy state. This isn't much different from atoms, electrons, and orbitals, where a low-energy state is preferred over a high-energy state.

Thermodynamics and Kinetics

Thermodynamics doesn't always have to do with anything actually moving but is instead a way of looking at how something might be more or less stable depending on its state. Kinetics is related to thermodynamics. It is used in chemistry to describe how fast a reaction happens or a substance changes over time.

For example, if you put two molecules in a beaker in a solution and wait 30 years, you would possibly see that molecule AB plus molecule CD turned into two separate molecules, called AC and BD. If you do the same thing and add an enzyme to the solution and the reaction happens in a few hours, you can say that the enzyme enhances the kinetics of the reaction. The two reactions are the same, but the speed of the reaction will increase by using the enzyme.

Let's look at a typical reaction diagram, which graphs the energy of a reaction over time. The starting molecules are the *reactants* and the ending molecules are the *products*. As long as the energy of the products is lower than the energy of the reactants, the reaction will happen and stay that way. There may be a bump in the middle that involves the activation energy necessary to get it all started but this will be temporary. This is what it looks like:

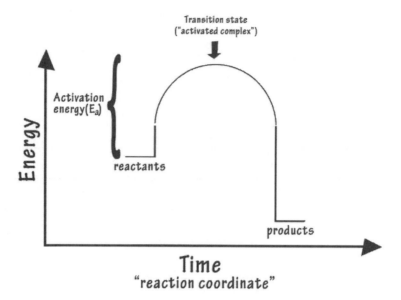

While kinetics is related to time, thermodynamics isn't. With thermodynamics, there are just two energy states: *reactant energy* and *product energy*. Any time the product energy is less than the reactant energy, the reaction between the two will want to happen. If not, it is the proverbial "uphill battle" to go from the reactant to the product in any stable way.

Just because a reaction is considered *thermodynamically favorable* doesn't mean it really will happen. Diamond and graphite are both forms of carbon, and a great deal of energy or pressure went into making diamonds out of graphite, but you will not see diamonds melting into graphite, even though its energy level is lower. This type of reaction would need a large amount of activation energy to turn a diamond back into graphite.

Any time you turn a molecule into another molecule, mix molecules chemically, or even turn diamonds into graphite, you will have to break a lot of chemical bonds. Sometimes, breaking these bonds isn't difficult and takes very little energy. Other times, the bonds are very difficult to break and the activation energy (which leads to an *activated complex*) is quite high.

The free energy change, which chemists call the *Gibbs free energy*, is the difference between the beginning *reactant energy state* and the *end product energy state*. If this Gibbs free energy change is negative, the reaction will happen for sure (but the time it takes to do this is not the same for every reaction, even if they have the same Gibbs free energy change).

We also say that this reaction will have a large *equilibrium constant*. This means that, at the end of the reaction, more products will be present than reactants. A small equilibrium constant means that the reaction will not give as much product as you'd like. Equilibrium constants, called *K*, do not apply to pure liquids or to solids. They can be calculated in a laboratory by measuring the concentration of products and reactants in a solution.

Enzymes in biological systems never change the beginning or ending energy levels but will change the activation energy so that it doesn't take as much energy to help the reaction happen. Enzymes are called *catalysts* because they speed up reactions. The whole idea of catalyzing something means to speed it up. Enzymes have active sites that allow one or more reactants to get close enough to one another in the right configuration to react. Once this happens, the products are released, and the enzyme is generally free to react to another set of reactants. This is what it looks like:

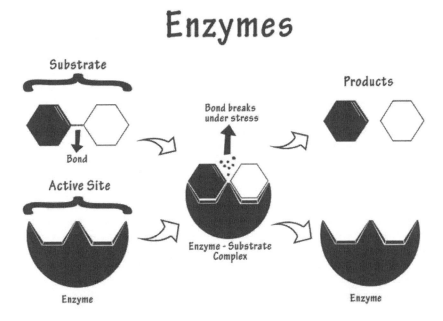

You see how enzymes can help a difficult reaction that would probably work anyway to occur more easily. It is not likely to help a reaction that is destined to go back to reactants anyway (because the energy level change will not be negative but will instead be positive).

Thermodynamics in Chemistry

Thermodynamics studied in chemistry involve concepts like whether or not a reaction will happen and different phase states of the elements and molecules. You need to know that, in chemistry (as in everything), energy goes from a high energy state to a low energy state without intervention. If, for example, you dump a bucket of water at the top of a hill, it will flow to a lower energy state down the hill. You can put energy into it, though, and carry the bucket of water up the hill — all at once or spoonful by spoonful.

A spontaneous chemical process is one that will naturally happen, even if no energy is put into it. A nonspontaneous process won't happen by itself unless energy is spent to drive it there. Many chemical processes are not all-or-none. This means that there will be some reactants left, even as there might be more products made during the reaction. If you see a reaction written like this, it means it is an incomplete one:

$$AB + CD \leftrightarrow AC + BD$$

Try to get a block of ice to melt on a hot day and you'll see spontaneous process. If you try to take a puddle of water on a hot day and freeze it by itself, you'll see a definite nonspontaneous process. It's the same thing that happens with certain chemical reactions.

Spontaneity does not equate to speed. A dead tree will spontaneously decompose but it could take months or years. Carbon-14 is used in carbon dating fossils because it spontaneously decays, but it takes a predictable rate to do this with a half-life of 5,730 years.

Gases in chemistry are important to biological systems and your cells. Carbon dioxide and oxygen diffuse in and out of the cell membrane of every cell, with many variables playing a role in how they do this. It depends on the difference in concentration of each gas across the membrane, the type of gas involved, and the temperature. Thermodynamics can explain these processes.

If there are two side-by-side systems that can't interact, their temperature will stay roughly the same and won't mix with one another. If they *can* interact, however, heat will flow from a hotter area to a colder area. Gases do the same thing. They flow, or *diffuse*, from a more concentrated area to a less concentrated area. This is what it looks like:

Heat Diffusion

$$T_X > T_Y$$

X and Y in contact

Gas Diffusion

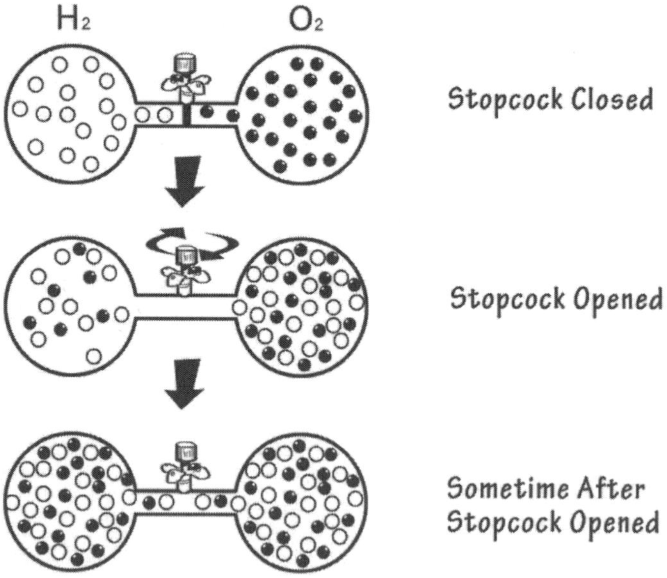

Stopcock Closed

Stopcock Opened

Sometime After
Stopcock Opened

Another term in thermodynamics you should know is called *enthalpy*. Enthalpy is related to both heat and energy. We measure heat by using a thermometer. A gas in a nonchanging space and at the same pressure will have the same enthalpy and *entropy*. If the pressure changes, though, the enthalpy will change, even though its heat won't change.

In a chemical reaction, the enthalpy might increase or be positive. It looks like this: $\Delta H > 0$. In this case, energy is being absorbed, and the reaction is called *endothermic*. It might look to you like the beaker where the reaction is taking place is getting colder. If the enthalpy decreases or is negative, it looks like this: ΔH<0. Energy is expelled from the system and the reaction is *exothermic*. You might feel the beaker getting hot instead as heat is given off. Endothermic reactions and exothermic reactions can both be spontaneous. It depends on whether or not there is energy available for the reaction to draw from.

In chemistry, there will be *reversible* and *irreversible* changes. If you put gas in a piston chamber and expand it using heat, this is reversible. If you look at the above image, you will see that gas raises the piston and the volume expands. If you cool the chamber, the piston will reverse itself and the chamber volume will decrease. This is a reversible condition.

If instead, you take a full balloon and heat it up, it will expand until it bursts. Gas will escape quickly, and the balloon will deflate. This is obviously an irreversible reaction because you can't possibly put all that air into the balloon. All of these things can happen in chemical systems, so some will be reversible and others will be irreversible.

Entropy, or *disorder*, also applies to chemistry. All reactions and phase changes in chemical systems can increase or decrease entropy. Entropy is defined by the letter S. If the change in S is positive, entropy will be increased. The reverse is true if the change in S is negative. You measure this by looking at the disorder of the end products and the reactants. The entropy depends not on what the substance looks like to the naked eye but on how disordered it is on a molecular or atomic level.

This idea of entropy makes the most sense when looking at changes in phase, such as when liquid water turns to water vapor or when looking at water vapor molecules as the temperature rises. If you think about the molecules of water in a liquid versus a gas, you can imagine that there is more disorder to gaseous water because they can bounce around to a greater degree and are more spread out than liquid water. This is what it looks like when water changes:

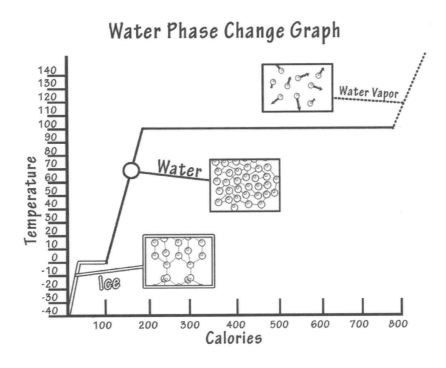

You see how the solid water has low entropy and how gaseous water is disordered with high entropy. The same thing happens with gas as you heat it. This image looks at how gas under higher temperature is more disordered:

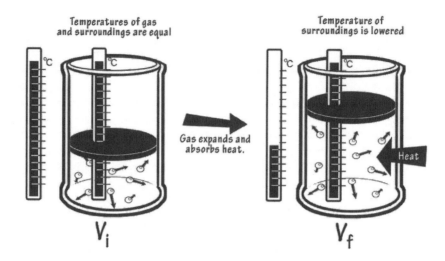

If you add heat energy to gaseous molecules, they bump against one another more readily. If the gas can expand under these conditions, it will. If it can't, the pressure of the gas will increase because these jittery molecules don't go anywhere.

What chemical reactions will increase or decrease the entropy of a system? Let's look at some examples:

- Increasing the volume of a gas (such as pulling on a piston of gas in a container) will increase entropy.
- Taking a solid and making it dissolve in a solution will increase the entropy of the solute/solid. You can sometimes, however, decrease the entropy of the solvent itself so that the total entropy of the system actually decreases.
- All phase changes from solid to liquid or from liquid to gas will increase the entropy of the substance undergoing that change.
- Dissolving a gas in a liquid will decrease its entropy. This is because the gas molecules are more widely distributed and move about to a greater degree as a gas but not so much as a liquid solute.
- All phase changes from liquid to solid (as in freezing) or from gas to liquid (as in condensation) will decrease the entropy of the system. Again, this makes sense when you look at molecules in the gaseous, liquid, and solid states.
- For any gaseous reaction that results in fewer total molecules (as in combining two gases into a larger gaseous molecule), the entropy will decrease as there is less disorder in the system. This is what it looks like:

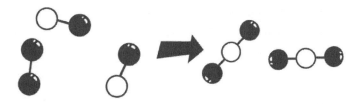

Any time there is less freedom of motion — being in a solid lattice or being bonded to another molecule instead of unbonded to it — it decreases the entropy of the system.

Motion Inside Molecules

There are different ways that molecules can move *within the molecule itself.* This image shows you a molecule and the different atomic movements possible inside the molecule:

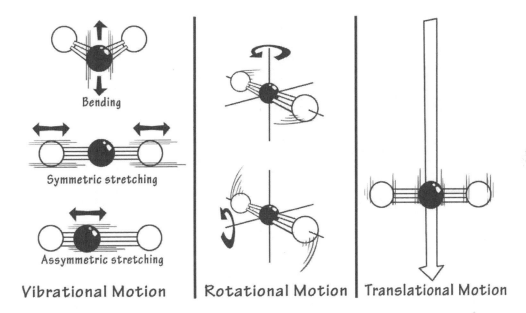

Translational motion is the motion of the whole molecule anywhere in space. *Rotational motion* is motion within the molecule around some type of axis. *Vibrational motion* is also within the molecule but involves the stretching or bending of the molecule. These actions are called *molecular freedoms of motion.* The more freedoms of motion in a molecule, the higher its entropy will be.

Gibbs Free Energy in Chemistry

The Gibbs free energy, or *G*, is also sometimes just called "free energy." It was named after J. Willard Gibbs, an American physicist in the late 1800s. It is related to the temperature, enthalpy, and entropy of a system. Just as in other systems, the change in Gibbs free energy in chemical systems determines the spontaneity of a reaction. This is what it looks like:

- If $\Delta G < 0$, the process occurs spontaneously.

- If $\Delta G = 0$, the system is at equilibrium.

- If $\Delta G > 0$, the process is not spontaneous as written but occurs spontaneously in the reverse direction.

So, the ΔG is the change in Gibbs free energy. This is what it looks like on a graph:

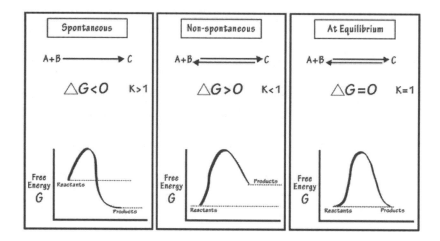

If you don't yet believe that energy is energy and that energy and work are created, let's look at a few examples:

- **Chemical battery** — This takes chemical energy and turns it into electrical energy with 90 percent efficiency. (The rest becomes the heat you feel from an operating battery.)
- **Home furnace** — This takes chemical energy from fuel and turns it into heat with about 65 percent efficiency.
- **Leaves from plants** — These take light energy and turn it into chemical energy, but the efficiency is just 30 percent.
- **Liver cells** — These take one kind of chemical energy and turn it into another kind of chemical energy but with a 30-50 percent efficiency rating.
- **Large electrical generator** — This takes mechanical energy and turns it into electrical energy with an efficiency of 99 percent.
- **Light bulb** — This takes electrical energy and turns into light energy with an efficiency of 5 percent for regular incandescent bulbs or 20 percent for a fluorescent bulb.

There is one more concept you should know about with regard to chemical reactions and enthalpy versus entropy. A favored state for entropy is one that is more disordered. A favored state for enthalpy is one that has the lowest energy state. A reaction could go either way. In the reaction, the letter "q" is heat that is either given off or put into the reaction. Which of these things happens depends on the direction the heat is moving.

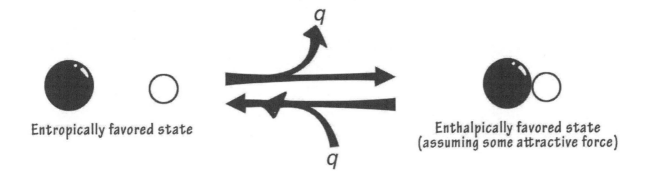

Entropically favored state   Enthalpically favored state
(assuming some attractive force)

To Sum Things Up

In this chapter, we looked at *thermodynamics.* Thermodynamics involves energy states and ways in which energy is transferred from one thing to another. Remember the laws of thermodynamics and understand that they apply to all systems in nature, from physics and machines to biological systems and biochemistry.

Remember what *enthalpy* and *entropy* mean. Entropy is a measure of disorder in a system. Without putting energy into a system in order to restore order, the tendency is to increased disorder. Enthalpy is related to heat energy. Chemical and molecular systems also prefer to be in the lowest possible energy state.

We will talk about enzymes more in another chapter. In biological systems, enzymes will lower the activation energy of a reaction but will not change the differences in energy states between the reactants and products in a biochemical reaction.

CHAPTER 11:
ATP AS "FUEL"

We've talked so much about ATP so far that you probably think you know everything there is to know about it. In this chapter, we're going to add to what you already know to give you a much better grasp of ATP.

ATP, or *adenosine triphosphate*, is the major energy currency in all cells. There are other molecules that do have the potential to put energy into a system but a lot of these just go onto making ATP anyway, so they really aren't used to drive any reactions besides those leading to ATP as the end product.

Chemically and biochemically, most reactions in the cell that need some type of energy input get this energy from the breakdown of ATP. ATP itself isn't like a lightbulb, where energy can be seen as light given off. Instead, it "carries" energy in one of its chemical bonds. When this bond is broken, the energy given off will affect enzymes or change proteins in some way where actual chemical work can be done.

ATP has a specific structure. It has an adenine base that is identical to the adenine seen in DNA and RNA molecules. In fact, ATP is used to make these nucleic acid polymers. It also has a ribose sugar attached to the 9-nitrogen atom on the adenine molecule. The 5' carbon sugar atom is where the three phosphate molecules are attached in a row. This is what it looks like from a chemical aspect:

The one thing you don't see on the image is that because ATP is negatively charged, it needs a cation to make it electrically neutral. Magnesium ions are the main cation that attach to the ATP molecule, with these ions attached to the negatively charged oxygen ions in the phosphate molecule.

How ATP is Made

We will talk a lot more in a later chapter about the metabolic processes that lead up to the creation of ATP in the cell. There is a long line of different cellular metabolic reaction steps that start with a molecule of fuel — like glucose or a fatty acid — and end with up to approximately 32 ATP molecules once a single molecule of these types of fuel is completely metabolized. How many molecules of ATP are made per cell and what the end products are depend on the cell and whether or not oxygen is present.

Briefly, these are the steps used to make ATP energy. In eukaryotic cells, these are clustered into biochemical pathways called *glycolysis*, the citric acid cycle (or *Krebs cycle*), and *oxidative phosphorylation*. When fatty acids are used as fuel, the process is called *beta oxidation*.

Glycolysis is so important that all cells in nature will use this process in some way. It all starts with the glucose monomer molecule. There are several biochemical pathways that first cost a little bit of ATP energy to get going but then, there is a "payoff" phase that will make more ATP molecules than were necessary in the beginning, plus a molecule called *NADH*, which is used later to make even more ATP energy. The end result is that there are two smaller molecules of pyruvate that can go onto making ATP energy through still other biochemical means. This is a summary of glycolysis:

There doesn't seem to be any oxygen as part of glycolysis; this is because it doesn't need any and explains why glycolysis is called *anaerobic respiration*. Even organisms that do not use oxygen participate in this biochemical process. It doesn't make a lot of ATP energy, and it doesn't fully oxidize, but it is a good thing for organisms that can't or don't have access to oxygen.

Fun Factoid: What do your muscle cells and beer have in common? It's maybe not what you'd think. In situations where you exercise so much that oxygen can't get to your muscles fast enough, your muscles undergo glycolysis to provide a fast energy source. The end result is that you build up a byproduct called lactic acid in your muscles (which makes them hurt later). The organism that makes beer (Saccharomyces cerevisiae) also undergoes glycolysis, but its end product is ethanol or alcohol, which is in the beer you drink. Both of these processes are called anaerobic respiration. *With beer, though, we also call it* fermentation. *See? They are related!*

The next part of metabolism to help make ATP energy and other energy molecules is called the *citric acid cycle*. It is sometimes also called the *Krebs cycle* or the *tricarboxylic acid cycle*. It is a real cycle in metabolism, where molecules of pyruvate are first turned into acetyl-CoA (giving off a carbon dioxide molecule and making an NADH molecule). This two-carbon acetyl group (attached to a molecule called coenzyme A) reacts with oxaloacetate already in the cycle to give a six-carbon molecule called citrate.

This figure simplifies the Krebs cycle according to the number of carbon atoms in each molecule. While only one molecule of ATP is made, it makes three molecules of NADH (per cycle) and one molecule of FADH2 (per cycle). You need to double that because there are two turns of the cycle necessary to burn up the entire glucose molecule into carbon dioxide. Glucose has six carbons and, by the end of the Krebs cycle, all six have become carbon dioxide.

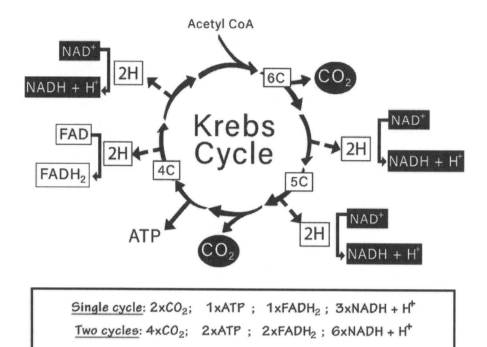

Single cycle: $2xCO_2$; $1xATP$; $1xFADH_2$; $3xNADH + H^+$

Two cycles: $4xCO_2$; $2xATP$; $2xFADH_2$; $6xNADH + H^+$

You won't see any oxygen used in the Krebs cycle itself but that doesn't mean it isn't necessary. The Krebs cycle is ultimately crucial to recycling the FADH2 and NADH made in this cycle. You would have to keep making precursors to these molecules and they would just build up if you didn't have oxygen to recycle them back to their precursors.

The final step in the process of making ATP (and the process of making the most ATP) is called *oxidative phosphorylation*, or the *electron transport chain*. You might think that nothing metabolic really happens there because the entire glucose molecule has been oxidized (broken up into carbon dioxide) but actually a great deal happens in this process.

NADH and FADH2 get recycled back to make NAD+ and FADH (which are "spent" energy molecules). These processes essentially cause an electrochemical change in the mitochondria of the cell that drives the enzyme that makes a lot of ATP molecules. This enzyme is called *ATP synthase*.

ATP Synthase

ATP synthase is the actual enzyme that makes ATP. How it does this is really amazing! As you will see, enzymes are not magical proteins that somehow get things done. Actual work happens in these molecules that is very similar to the work done on a much large scale in motors and engines.

ATP is an enzyme that is partly imbedded in the inner membrane of mitochondria. It has many parts. The part hidden within the membrane is called the F0 part. It is an electrical motor that does real work. How is it that an electrical motor can exist in your cells? Let's look more closely at this enzyme to see how it works.

The F0 segment works because there is an electrical charge across this membrane. This electrical charge is called the *proton motive force.* It'd made by all of the FADH2 and NADH created earlier in pathways like glycolysis and the Krebs cycle. The proton motive force is basically protons (charged H+ atoms) located on one side of the membrane but not on the other.

The protons flow through the membrane in order to try to equalize the charge. This turns the F0 circular rotor part of the enzyme, which is connected to the F1 part, a chemical motor. These two parts are connected to one another through a part called the *stator* so that when F0 rotates, the F1 part rotates as well. Here's what these look like:

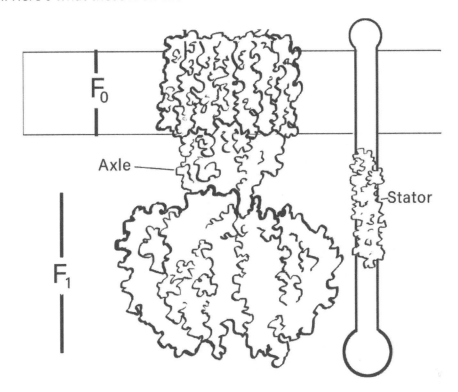

So, now what happens? These two motors together make a generator. The F0 segment rotates using this proton motive force and the F1 segment rotates as well, generating ATP molecules. The axle also turns (like the axle of your car wheels) to turn both motors. The F1 part attaches an ADP molecule (which is an ATP molecule but with two phosphate groups) plus a phosphate group to make ATP. Once the ATP is made, it drops off and another ADP molecule starts the process over.

Your cells do not make ATP from scratch very often — mostly because they do not need to. Once ATP is used up, it becomes *ADP.* This just gets recycled back into ATP, provided there is energy to do this. The total amount of ADP and ATP together in your cells remains the same most of the time.

You do amazing work with the ATP in your cells, hydrolyzing it to make ADP and phosphate so many times that the 0.2 moles of ATP in your entire body get recycled hundreds of times a day. About 100 to 150 moles of ATP get hydrolyzed in your entire body each day — amounting to about 50 to 75 kilograms per day (this is basically your entire body weight recycled every day). It is estimated that one ATP molecule gets recycled up to 1500 times per day.

What Does ATP do Inside Your Body?

ATP does many things inside and outside the cells of your body. It is a good signaling molecule (which is a molecule that sends a message to another molecule in the cell). There are enzymes called *kinases* that use ATP all the time. Kinases are enzymes whose function it is to add phosphate molecules to other molecules. There are a lot of kinases in your cells. They have a binding site for ATP and use the phosphate group given off when ATP hydrolyzes into ADP in order to give that phosphate group to another molecule. Many cellular processes depend on kinase enzymes and their ability to phosphorylate (add phosphate groups) to molecules.

The opposite of a kinase is a *phosphatase enzyme.* In this image, you can see how a protein kinase (PK) adds a phosphate group using ATP and how a protein phosphatase (*PP*) takes a phosphate group off.

In some cases, the substrate gets "activated" with a phosphate group attached, while in others, it gets "deactivated."

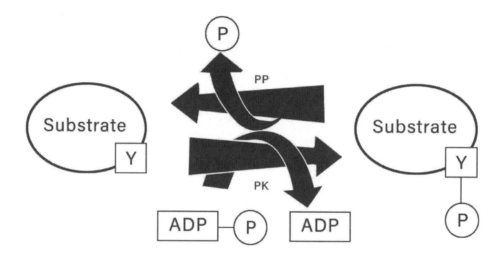

ATP is essentially a monomer unit that helps make the DNA or RNA polymer. You can imagine that it takes the conversion of ATP (which has a ribose sugar attached) to deoxyribonucleotide ATP (dATP) in order to make the DNA polymer. ATP also activates part of the process in the ribosomes of the cell that add amino acids one at a time to make protein polymers.

There are two common cellular functions that use ATP as a crucial part of how they work. One of these is called the *sodium-potassium ATPase pump,* which is a pumping enzyme/channel in the cell membrane. Its job is to pump sodium out of the cell and pump potassium into the cell. It does this all the time in every cell, using up about 20 percent of all the ATP you make. The idea behind it is to set up a difference in electrical charge across the membrane. How does it work?

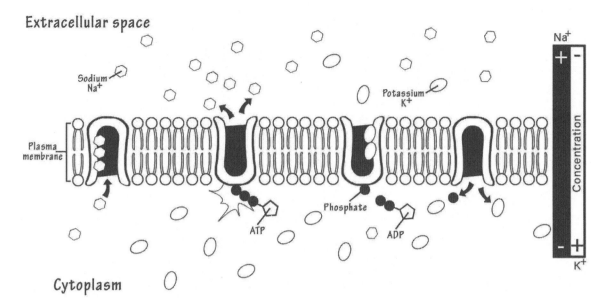

You can see how important ATP is just by looking at the different steps involved in this pump's activities. These are the steps:

1. The enzyme starts out open on the inside where it loads three sodium ions into the channel protein.
2. ATP gets hydrolyzed to make ADP on the inside of the cell.
3. This causes the carrier channel protein to change shape and to drop off the three sodium ions into the space outside the cell.
4. The channel protein is now open to the outside and has a strong affinity for potassium ions on the outside of the cell. It loads up two of these.
5. The potassium in the channel protein changes the phosphate on the inside of the cell (that was first attached by the ATP molecule). This reconfigures the channel proteins shape again.
6. The new channel protein shape opens to the inside, discharging the two potassium ions, resetting the system to start again.

If you have paid attention to your math in this process, you will notice that three sodium ions (positively charged) exit the cell but only two potassium ions (also positively charged) enter the cell. This means there will be more positive electrically on the outside and less on the inside. This negative charge on the outside compared to the inside is called the *membrane potential.* It takes a continual source of ATP energy in order to sustain this membrane potential on a 24/7 basis.

Another big job for ATP relates to the way your muscles contract. ATP helps muscles cells get the signal to contract but it also helps with the actual contraction process in a unique way. Before we discuss how ATP is essential for this process, let's look at what makes a muscle cell contract in the first place.

Muscle cells have long protein filaments in them; they are made from a thin *actin filament* and a thicker *myosin filament.* A bunch of these in a bundle of skeletal muscle forms a unit called a *sarcomere.* One whole muscle cell has a long line of sarcomeres attached end-to-end. A sarcomere looks like this:

Sarcomere

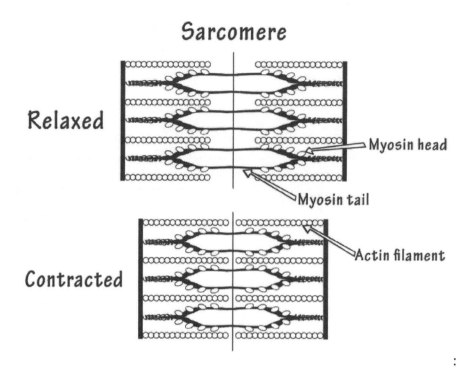

Relaxed

Myosin head

Myosin tail

Actin filament

Contracted

Your muscle contracts because the actin filaments crawl along the myosin filaments using bridges called *cross-bridges* between them. It's a lot like a ratchet wrench that slowly jerks the filaments past each other segment-by-segment. It's really not too complex but it does require ATP energy.

The whole thing cannot happen without cross-bridging between the two filament types. Myosin is not a smooth fiber; it has lumps every so often along its length. These lumps are globular proteins with an S1 section that acts like a hinge. When the crossbridge happens, the S1 region bends in a flexible way in order to "walk" the two fibers so that the sarcomere shortens. This is what it looks like:

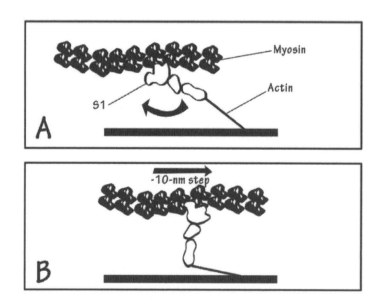

This bending of the S1 segment is called a *power stroke* and requires ATP in order for it to happen. This image shows the power stroke in the different steps:

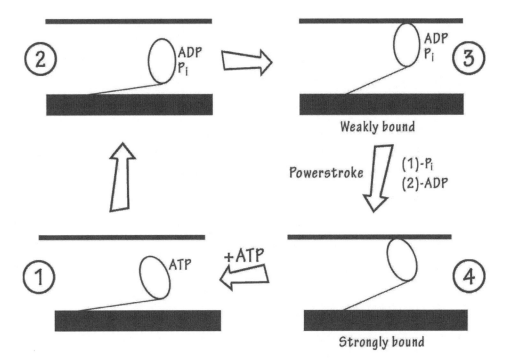

The process is a little more involved than this, but you should know that, while the attachment of myosin to actin using the crossbridge is relatively easy and does not involve ATP, the power stroke and the re-cocking of the bridge to walk the two filaments in such a way as to have a muscle contraction does need ATP energy.

To Sum Things Up

ATP is the "energy currency" of the cell. The entirety of its energy is held within the third phosphate bond so that, when ATP is hydrolyzed to make ADP, there is energy given off as part of this process in order to allow enzymes to do their job more effectively.

ATP is made by burning fuel. The fuel most commonly used is *glucose,* which can be oxidized to make carbon dioxide and water. As it does this, ATP can be made directly. It can also be made by making other molecules, namely FADH2 and NADH, which create a proton motive force across a membrane inside the mitochondria of the cell. This provides the energy to rotate molecules that effectively make ATP molecules in the end.

ATP has many jobs. It can activate and deactivate proteins and enzymes. It runs the sodium-potassium ATPase pump necessary to have a charge differential across cell membranes. It also helps things like muscle contractions through driving the power stroke that allows the different protein chains in the muscle cell to walk along each other as part of the contraction of each cell.

CHAPTER 12:
METABOLISM AND ENZYMES IN THE CELL

When you think of metabolism, you probably think about burning fat or expending calories. While these are both involved in metabolism, the truth is broader than that. Metabolism involves any chemical process that either breaks down or builds a new molecule. So, it goes both ways. It also involves activities on a very large scale as well as activities on a molecular level inside the cell.

You can divide metabolism into two categories. These are called *catabolism* and *anabolism*. Catabolism or catabolic processes are what you think of when you talk about burning fat. These are all processes that break down a molecule into smaller parts. We've talked about glycolysis and the Krebs cycle. Together these are catabolic pathways that turn glucose into six carbon dioxide molecules per molecule of glucose you started with.

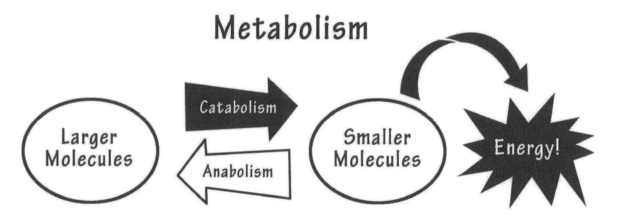

On a large scale in your body, catabolism happens when you eat food that gets broken down into smaller monomers in order to be absorbed by the GI tract. If you eat a piece of steak, you chew it and chemically digest it so that it becomes amino acids. You absorb your steak as amino acids rather than as a chunk of meat protein.

This is basically how your GI tract works:

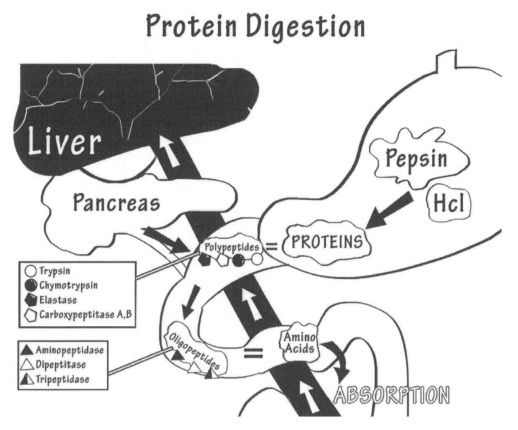

While catabolism produces energy, anabolism needs energy. Anabolic processes include those that build proteins, enzymes, carbohydrates, and lipids. You will see how these processes work in the cell to build all of the structural and functional components.

In your body, not all of the energy given off by catabolic processes goes into all the anabolic processes inside your cells. The rest is given off as heat (which is why you are hot-blooded) or is stored as fat in your fat cells (or as glycogen in your liver and muscle cells). Anabolic and catabolic processes go together intimately, so you can't have one without the other. These form your metabolism.

How Catabolism Works

If a reaction takes a large organic molecule and breaks it into smaller molecules, you can be sure it is a catabolic reaction of some kind. While these work well to make energy for the cell, they aren't completely efficient. About 40 percent of the energy made actually goes into making ATP that will immediately be available for use by your cells (almost always to be used for anabolic processes). A small amount of ATP can be stored but it is so reactive that it is difficult. The rest (nearly 60 percent) goes to making heat, which your cells absorb to keep them warm.

This is your metabolism in a nutshell: Your food is eaten and broken down before getting absorbed. The breakdown products (monomer sugars, fatty acids, and amino acids) either get metabolized further (or you could just say *catabolized*). Some of these products will go on to making structural components in the cell as shown:

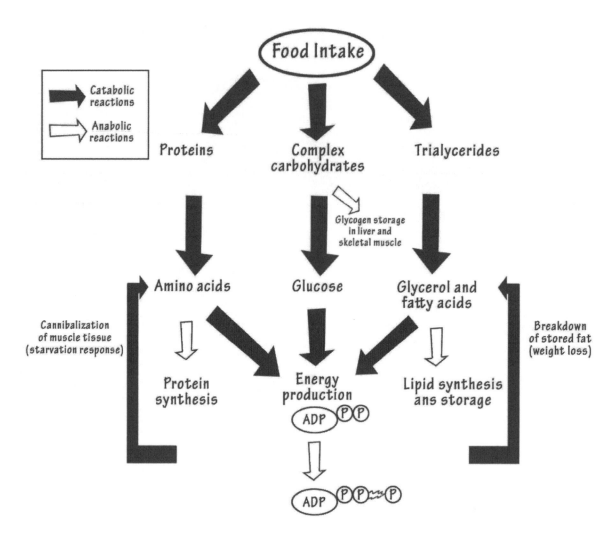

As a fuel source alone, glucose and other simple sugars are the main sources of the energy you need for the cell. You need to eat your carbs, especially for your brain, which can only use glucose as its energy source.

Your body tightly controls glucose levels and your metabolism through hormones like *insulin* and *glucagon*, which are both made by the pancreas. Without insulin, the sugar in your bloodstream won't enter the cells and blood sugar will increase. This disorder, called *diabetes*, means your cells must use other energy sources, such as fatty acids and amino acids, for fuel.

With a lot of sugar and insulin present, the liver will be able to store glucose as glycogen in your muscle and liver cells. Anything leftover goes to making triglycerides/fat in your fat cells. This is ideal and works perfectly most of the time but, if you eat too many calories (more than you burn), you will store too much as fat.

Fun Factoid: A diabetic who doesn't know he or she has it and gets severely high blood sugar will often lose weight and pee a lot. Why do these things happen? First, even though there is a lot of fuel present, the fuel injector (insulin) isn't working, so no glucose can get into the cells. They starve and don't grow, so the diabetic will lose weight. Second, the extra glucose must be gotten rid of somehow, so the kidneys filter it out into the urine. Where glucose goes, water must follow, so the diabetic will often be thirsty or dehydrated, even though they pee a lot!

Lipids you eat become fatty acids; these will undergo *beta-oxidation* (another metabolic pathway just for fatty acids) in order to be used to make ATP energy for the cells. If there is excess eaten, it will go into making triglycerides that are stored in the fat cells. Half of your fat protects you from the cold and is found just under the skin in the *subcutaneous tissue*. The other half is called *visceral fat* and is located around your organs and your waist. This is the fatty tissue most linked to diseases like diabetes and heart disease (not so much your subcutaneous fat).

Proteins are least likely to be used as fuel. These are only used if you are chronically starved and have no other fuel source left. Otherwise, the preferred pathway for proteins is to be broken into amino acids and then directly put into making other proteins in the cell.

Nucleic acids are eaten too, but we don't often think of these as being part of nutrition. They get broken into nucleotides in the GI tract and then get absorbed. They often go directly to the cells of your body in order to be made into nucleic acids all over again for your own body's cells.

Regulation of Metabolism

We talked about how insulin helps to regulate glucose metabolism, but it isn't the only hormone by far that is involved in regulating metabolic processes. There are many different catabolic hormones you need to break down substances for fuel. These include *epinephrine* (which speeds metabolism by activating your autonomic nervous system), *cortisol, cytokines,* and *glucagon* (made by the pancreas).

There are also anabolic hormones that help enhance biosynthesis, or *anabolism*. These include *insulin, growth hormone, testosterone, insulin-like growth factor,* and *estrogen.* Testosterone, for example, stimulates protein production, which explains why men (who have a lot of testosterone compared to women) are so beefy sometimes. Women just don't have the testosterone for major muscle accumulation, although they still can build muscle through exercising.

Fun Factoid: *What about anabolic steroids? Are they good for athletes or not? These are hormones that mimic testosterone in the body, so they are sometimes taken by body builders and other athletes to "bulk up." They will do this, but the side effects can be terrible. Men can grow breasts, have shrunken testicles, and a low sperm count. Women who take it will have lower voices and appear more masculine. Besides benign things, like acne and male-pattern baldness, anabolic steroids can cause fatal liver cancer, heart attacks, and strokes. In the end, they might not be worth it.*

Enzymes and What They Do

We've talked a lot about how enzymes work when we discussed thermodynamics in living systems. Enzymes lower the activation energy of a biochemical reaction so that they happen more easily and with faster kinetics. There are a few enzymes made out of RNA molecules but most by far are made from globular proteins. Remember, they can't force a reaction that wouldn't otherwise happen.

The enzymes of your body are very sensitive to their environment. If the pH is off (too high or too low), it affects the protein shape and will affect enzyme activity. It is also sensitive to temperature. If the temperature is too low, reactions for all biochemical processes will be slower. If the temperature is too high, enzymes can break down altogether or simply change shape, so they won't work as well.

This is an energy diagram showing what happens with and without enzymes being involved in the reaction process. Again, the energy levels in the beginning and at the end of the reaction will be exactly the same; just the activation energy changes.

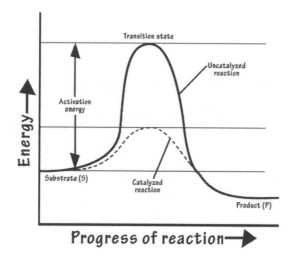

In a chemical reaction using enzymes, the reaction looks like this:

$$S + E \rightleftharpoons ES \rightleftharpoons E + P$$

S is the *substrate* and *P* is the *end product*. The enzyme *E* will temporarily bind to the substrate, alter it in some way, then turn it into a product. The enzyme itself stays the same so that it can be continuously recycled over time. It doesn't take much enzyme to cause a lot of substrates to go into making products.

Let's look again at how an enzyme works. It looks a lot like this when a biochemical reaction occurs:

Enzyme-substrate complex

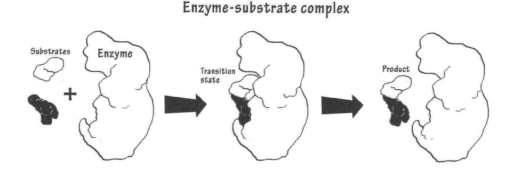

One interesting trick is that once the substrates react along with the enzyme, the product doesn't fit as well into the active binding site of the enzyme, so it detaches relatively easily and leaves the enzyme's vicinity. This readies the enzyme to start all over again. Part of how many enzymes work is to get two substrates close enough together and in the right orientation so that they can easily react — much easier than they otherwise could without the enzyme involved.

Sometimes, you'll hear that substrates and enzyme active binding sites fit together like a lock and key. This is sort of true, but it is more like there is a good fit for the substrates and the binding site — called the *induced fit model* of enzymes. Once the reaction happens, the "fit" just isn't as good, so the end products will drop off much more easily than is true of the substrates. Isn't that interesting? It's all about how molecules fit together and react with one another, only on a massive scale with thousands of enzymes doing your cells' bidding all the time.

The substrates and enzymes might fit together because they are both hydrophobic or because they bind loosely through *hydrogen bonding*, *van der Waals bonding*, or *ionic bonding*. It will hardly ever involve any covalent bonding because the connection between substrates, products, and enzymes must be weak enough to allow the molecules to come together and break apart when necessary without a lot of chemical effort.

Sometimes, enzymes use other small molecules in order to help a reaction proceed favorably. These small molecules are called *coenzymes.* Some coenzymes bind to the active site along with the substrate, while others bind elsewhere in the system in order to make the enzyme have the best possible shape. Still others are needed for the enzymatic reaction itself (between the substrates). Many cofactors are *metal ions,* such as *iron* or *zinc.* Magnesium and other ions might act as enzyme cofactors. These cofactors also do not get used up but are recycled along with the enzyme.

NAD+ (which is like NADH but without the hydrogen ion added to it) is a common cofactor in *redox* reactions. NAD+ accepts a hydrogen ion easily and pulls it off a substrate (along with two electrons) to make NADH. This same NADH molecule can then go on to donate the hydrogen ion and electrons to another molecule in the same or a related biochemical reaction process.

Oxidation and reduction reactions are an important part of enzymes and their participation in cell metabolism. Together, these are called *redox* reactions because they always go together. In general, oxidation often involves oxygen and is involved in catabolism while reduction is involved in anabolism. You shouldn't think of these reactions this way, however.

A reduction reaction involves gaining an electron, while oxidation involves the loss of an electron. When oxidation happens and an electron is donated to some other molecule, energy is released. This molecule with the added electron is called the *reduced molecule*. It can go on even further to oxidize yet another molecule by passing its electron on. You can see how this would keep going until the electron ends up in a stable molecule that leaves the system.

While it seems contradictory to say that something is "reduced" by gaining an electron, especially since they also gain hydrogen ions along with the electron, it has to do with something called the *oxidation state* of an atom or molecule. The oxidation state is the same as what its charge would be if it were an ion.

Adding an electron would make an atom more negative or reduced to a lower oxidation state. It would be negatively charged if it were an ion. Removing an electron has the opposite effect, so it isn't as reduced to a negative charge state as it otherwise would be. It's complicated, but this is how it has been set up. Incidentally, a lot of oxidation in chemistry has nothing to do with oxygen — just to this *oxidation state*. This is what redox reactions look like together:

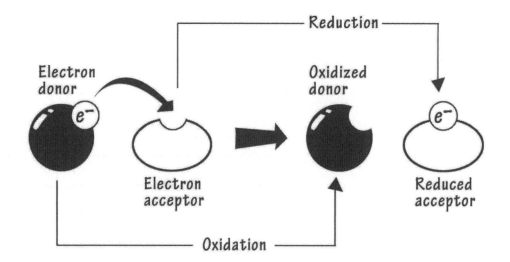

In biological systems, many redox reactions involve molecules like NAD+ and FAD. These are the oxidized forms of their respective molecules. When they take on electrons and hydrogen atoms, they are named NADH and FADH2, which are fully reduced. You'll see how this works later when we talk about metabolic processes in detail.

Coenzymes can also be vitamins in your diet; more likely, though, they are made from vitamins. One coenzyme you need is called *thiamine pyrophosphate,* which comes from vitamin B1 or thiamine. NAD+ is made from niacin, which is another B vitamin. Biotin is a vitamin that is itself a coenzyme. Folate can become *tetrahydrofolate*, which is yet another coenzyme.

Enzymes get regulated in several ways. They don't just act like runaway engines but instead have "brakes" on their activity. One way this happens is called *feedback inhibition.* It means that the product of a biochemical pathway "feeds back" onto other enzymes in order to prevent these enzymes from making more of the end product. Once the cell system says, "There is enough of this product," the process shuts down until more is necessary at a later time.

Feedback inhibition can happen in unique ways. One of these is called *allosteric regulation*. It means that the inhibitor (usually an end product of some kind) won't bind to the active site but will bind elsewhere to a site called the *allosteric site*. This could change the enzyme's shape in such a way that it simply won't work as well — even though the inhibitor molecule doesn't directly block the enzyme's active site. There are other regulatory molecules having the opposite effect on the enzyme — activating the enzyme rather than inhibiting it.

Remember when we talked about kinases and how they add phosphate groups to molecules? There are ways to regulate an enzyme by having a kinase nearby that phosphorylates it. This could turn on the enzyme or turn it off. It depends entirely on the system involved. *Phosphorylating* an enzyme is probably the most common way to change its activity but you can imagine that there are other systems that could add a different covalently bonded side chain to an enzyme in order to activate or deactivate it.

Types of Enzymes

We talked about *kinases* and *phosphatases.* You know from the enzyme's name that, if it is a kinase, it adds a phosphate group to a molecule. If it is a phosphatase instead, it will remove a phosphate from a molecule. It turns out that there are other tricks you can use to decide what an enzyme does just by looking at its name. These are the ones you should know about:

- **Oxidase** — These enzymes oxidize something by removing electrons from it as part of a redox reaction situation. It takes the electrons (and hydrogen ion) from an electron donor and gives it to another molecule. Sometimes these are called *dehydrogenases* because they remove hydrogen ions. An oxidoreductase enzyme will oxidize something while reducing another.
- **Transferase** — These are enzymes that transfer side chains or other molecules from one molecule to another. *Hexokinase* is also a transferase because it transfers a phosphate group from ATP to glucose.
- **Hydrolase** — These enzymes will transfer some type of molecule or side chain to water. These are technically transferases that just transfer the side chain to water as the acceptor molecule.
- **Lyase** — These are enzymes that either break double bonds or form them in a carbon-containing molecule. There will be two carbon atoms that have some side atoms removed in order to create a double bond between them.
- **Isomerase** — These are enzymes that simply rearrange a molecule. The basic chemical formula remains the same, but the atoms are switched around. These are sometimes called *oxidoreductases* within the same molecule because electrons also get transferred from one place to the next.
- **Ligase** — This enzyme is used to combine two molecules or parts of molecules using covalent bonding. A carbon-carbon bond could be made or a carbon-sulfur bond, among others. These are also called *condensation reactions* because they condense things together.
- **Translocase** — This involves an enzyme that will move a molecule or enzyme across some type of membrane. These are often located near or within the membranes themselves.

To Sum Things Up

Metabolism is a two-way street. In *catabolism*, larger molecules are broken down, often creating energy in the process. In *anabolism*, small molecules come together to make larger molecules, or *polymers.* These reactions often require energy to help them happen.

Enzymes are the main "helper molecules" in living systems. They allow reactions to be kinetically faster by reducing the activation energy of the biochemical reaction. These are often regulated by feedback inhibition, where the end products feed back onto the enzymes needed to make it in order to prevent more of the end products from being made.

DIVISION THREE:
CELLS AND WHAT THEY DO

SECTION FOUR:
INTRODUCE YOURSELF TO YOUR CELLS

You have come a long way so far in learning the lingo of physics, chemistry, and biochemistry. These things were important for you to learn before diving into the real study of cell biology. This is because cells are nothing more than intricate biochemical factories that use the principles of physics and chemistry — as is true of everything in the universe. Cells are the universe explained on a teeny tiny level. If you master the concepts in this section, you will be able to apply this knowledge to all life forms, large or small.

You are a whole multicellular being, but your entire body functions at an atomic level all the time. Atoms form molecules. Molecules interact to build cells and run biological processes in each cell. Each of your cells is unique but none operates independently from the others. As you study what happens in the cell, think about how this information is shared with other cells in your body so that the entire body is still a biochemical factory. Think of it as a super-factory system, a grand biochemical process on a very large scale. Read on as we study the cells!

CHAPTER 13:
THE DIFFERENCE BETWEEN
PROKARYOTIC AND EUKARYOTIC CELLS

The most basic cell type is the *prokaryotic cell*. These are tiny cells that are extremely important in the ecology and biochemistry of the world. *Prokaryotes* are basically just the bacteria and archaea organisms among living systems. They are always *single-celled organisms*.

Researchers have looked at how many prokaryotic cells there are on Earth. They came up with an estimate of 4 to 6 x 10^{30} prokaryotic cells. This is four to six quintillion cells, or about 5,000,000,000,000,000,000,000,000,000,000 cells.

Another estimator came up with a number of eukaryotic cells as being on the order of 10^7, or about 10,000,000 of these cells on our planet. This means that, on a cell-by-cell basis, prokaryotes outnumber eukaryotes by a factor of more than 100,000,000,000,000,000,000,000 to 1.

If you just measure the mass of carbon atoms within prokaryotic cells on Earth, you'd get 350 to 550 peta-grams of carbon. One peta-gram is 10^{15} grams; you can see that this is a great deal of carbon just within these tiny prokaryotic cells. They also have more than 10 times the amount of phosphorus and nitrogen compared to plants, which themselves have more of these elements than animals do.

Prokaryotic cells also divide rapidly, creating almost 2 x 10^{30} new cells. Most of these organisms live and divide in the open oceans, but you should know that prokaryotes are *everywhere*... even in the outer parts of our atmosphere and in the most hostile environments on earth. This all leads to a very high number and variety of these types of organisms on this planet.

Prokaryotic Cells Explained

Prokaryotes were the first cells on Earth. We will talk about how they evolved in a little bit but for now you should know that these were all we had on Earth for the first two billion years of life's existence. While eukaryotes are considered more evolved by far compared to prokaryotes, there are still way more of these tiny cells than there are of eukaryotes. In fact, if you take a handful of dirt and count the prokaryotes, you'd find more of these cells than the number of humans that have ever lived.

The two main types of prokaryotes are *bacteria* and *archaea*. (These are two main domains if you'll remember.) They are similar in that they look roughly like one another and are both types of prokaryotes, but archaea have biochemistry within them that are kind of a cross between bacteria and eukaryotes. This makes them a unique subset of prokaryotes different from bacteria in ways that you can't see with the naked eye.

This is a typical bacterial cell:

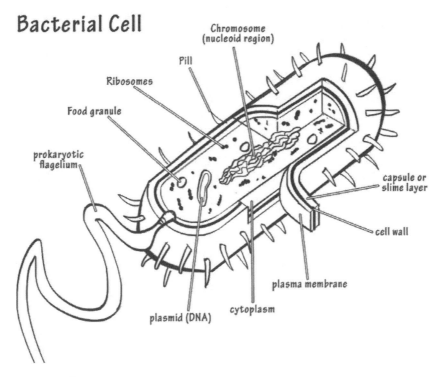

Bacterial Cell

- Chromosome (nucleoid region)
- Pill
- Ribosomes
- Food granule
- prokaryotic flagelium
- capsule or slime layer
- cell wall
- plasma membrane
- cytoplasm
- plasmid (DNA)

This is a typical archaea cell:

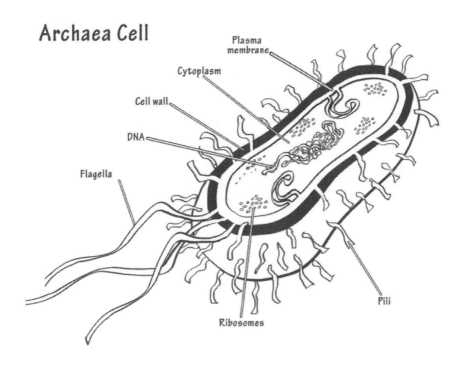

Archaea Cell

- Plasma membrane
- Cytoplasm
- Cell wall
- DNA
- Flagella
- Pili
- Ribosomes

If you can't tell the difference, it isn't because you haven't looked hard enough. As we will talk about soon, the major differences just cannot easily be seen (or can't be seen at all).

Prokaryotic Cell Structure

Prokaryotic cells are essentially empty structurally. They have a cell membrane surrounded by a cell wall. Inside, they have a nucleoid, which isn't a membrane-bound structure. It's just the area where the prokaryotic cell clusters its DNA. In the cytoplasm, you'll find many freely floating protein-making ribosomes — many of them clustered near the nucleoid. Other than these contents, the inside of these cells is quite boring.

The outside of prokaryotic cells is much different, however. They have interesting cell walls, and some have "capsules" around this. A few also have cilia and flagella in order to have the ability to move. We will talk more about how these structures work in prokaryotic cells. You will see that these cilia and flagella are different from those found in eukaryotic cells.

You may have heard that prokaryotes do not have organelles. This is true, but it isn't true that they don't have any internal membranes at all. A few specialized prokaryotes that participate in photosynthesis or in some type of cell respiration beyond glycolysis actually have a loose network of membranes that help perform these processes. Also, because the nucleoid has no membrane, it isn't technically an organelle either.

The parts of a prokaryotic cell are relatively simple. They all have an outer cell membrane, similar to eukaryotes but without components like cholesterol seen in animal cells. Cell membranes, as you will soon learn, are floating rafts of *phospholipids* mixed with proteins and other lipids. The phospholipid part keeps the inside of the cell separate from the outside of the cell — a very necessary function. The proteins do other things, like make channels for molecules to get inside or outside the cell or create a receptor or molecular signal outside the cell.

Bacteria and archaea also have cell walls. These are interesting coatings outside of the cell membrane that help the cell have structure, among other attributes. They also protect the cell from bursting or shrinking too much in certain hostile environments. The cell wall also contains some protein/sugar combination molecules called *glycoproteins* that tag the cell as being of one or another type.

The cell wall of prokaryotes is made mostly of giant polymers of sugars and peptides arranged in certain ways. In most bacterial organisms, this cell wall is made of *peptidoglycan*. This is what peptidoglycan looks like:

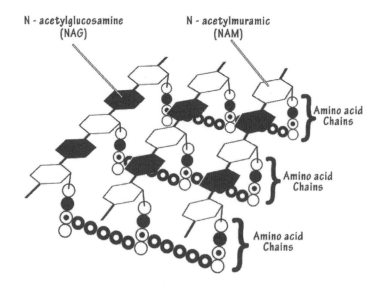

The two sugars are called *NAG* (which stands for *N-acetylglucosamine*) and *NAM* (which stands for *N-acetylmuramic acid*). These are essentially modified glucose molecules. You can see how these are connected by short peptide links that make an interwoven mat surrounding the cell membrane and protecting it. Some bacteria have thick cell walls while others have thin ones. Archaea have cell walls made from different sugars — most commonly a sugar called *pseudomurein*.

Outside the cell itself, there might be one or more flagella or multiple tiny cilia. These look different from one another but have similar purposes. A flagellum is a whip-like structure that moves in order to help the cell move. Cilia act together like tiny oars on a boat that move in unison to allow the cell to move in ways similar to flagella.

The bacterial flagellum is made from a long protein called *flagellin*, which is shaped like a helix. It has a sharp bend in it just as it exits the cell membrane so that when spun by a tiny motor in the cell, it spins around. Without the bend, it wouldn't move the cell in different directions. The motor is a large protein-based engine called a *Mot complex* that uses the energy you can get from a proton motive force in order to drive the rotary motor. (Remember that the proton motive force is just hydrogen ions separated by a membrane so that the movement of these ions across the membrane creates usable energy.) This is what a bacterial flagellum looks like:

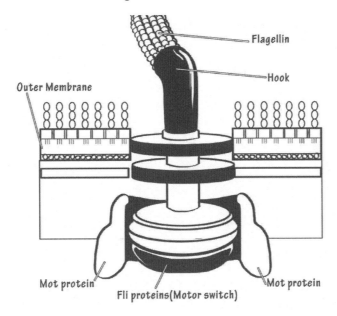

Archaea also have flagella but, while these look similar, they are so different in what they are made from and what drives them that they are considered completely unique structures.

Cytoplasm is the goo inside the cell. Most of it is water but it is so full of ions, sugars, enzymes, and other molecules that the substance is more gel-like than watery. This is where all of the molecules float about in the cell, diffusing from one part of the cell to the other. The environment inside the cell cytoplasm is tightly controlled by the cell membrane so that the enzymes are sitting in the ideal environment for proper functioning of the cell.

Ribosomes are small protein and RNA-containing structures that make proteins. In eukaryotes, a few of these float in the cytoplasm, but in prokaryotes, they have no other choice. These are made from two subunits, labeled the *50S* and *30S.* Together, they make a complex called the *70S ribosome.*

Bacteria have their DNA clustered in a single circular chromosome that is bundled up in a ball called a *nucleoid*. It has no membrane, but it keeps the DNA in one place. Ribosomes often cluster around it so that the DNA message inside the bacterial chromosome can easily be passed to the ribosomes in order to make proteins.

In many cases, the process of taking the DNA message and turning into a protein happens all in one fluid step. A segment of DNA could be transcribed or "read" into messenger RNA, while at the same time this same messenger RNA molecule can get translated into a protein molecule even before the messenger RNA molecule is finished getting made.

Bacterial Shapes

Bacteria come in several different shapes. The shapes don't really mean much or say much about the bacterial organism itself, but many bacteria have such a unique shape that researchers and lab technicians can make a good guess about what kind of bacteria they are seeing just by looking at the overall shape and how the different bacteria like to cluster themselves.

The different shapes are fairly easy to predict by their name. The *cocci* or *coccus* are bacteria that are circular or oval in shape. The *bacilli* or *bacillus* are cylindrical or relatively long and skinny. The spiral bacteria have a spiral shape, and the vibrio bacteria are comma-shaped. There are in-between shapes like the *coccobacilli* that are not complete cocci and not completely bacilli. This is what they look like:

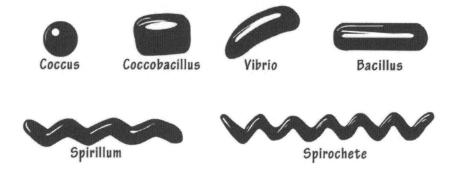

Bacteria and archaea both divide asexually through a process called *binary fission*. This is a simple process involving growth of the prokaryote's size, replication of its DNA to make two separate circular chromosomes, and division of the cell into two daughter cells. The daughter cells will be essentially the same as the parent cell.

There are a few ways that bacteria can take up new pieces of DNA in order to change how they look or what they do biochemically. In *bacterial conjugation*, one cell connects to a neighboring cell, sending a piece of DNA through the bridge between them. In *transformation*, the bacterial cell just sucks up some DNA from the environment, turning this piece of DNA into a working part of its own genetics.

There is also a process called *transduction*. This means that the DNA added to the bacterial cell has gotten in there using some kind of *vector*. Say, for example, that a bacterial virus (called a *bacteriophage*) grabs a piece of DNA from one cell before killing it and going onto the next cell. When it infects the next cell, this piece of tagalong DNA goes with it, essentially transferring the DNA from one cell to another.

All of these processes are called *horizontal gene transfer*. It's called horizontal because it doesn't involve different generations of bacteria but happens within the same generation. The piece of DNA can stuff itself into the circular chromosome of the new cell so it can be forever transferred to all subsequent daughter cells. It can also form a separate tiny circle called a *plasmid* that must divide itself and separate into the two daughter cells equally in order to get from one generation to the next. These are examples of horizontal gene transfer:

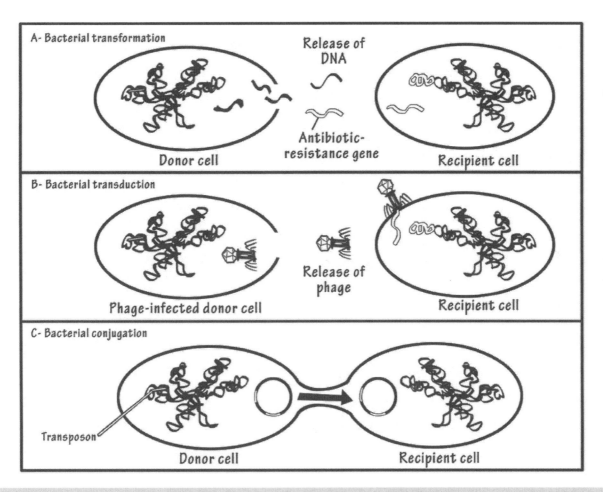

Fun Factoid: *It's through horizontal gene transfer that a lot of antibiotic resistance occurs. If just one cell mutates in an infection in such a way that it resists being killed by an antibiotic, this cell has its own evolutionary advantage, taking the place of less-fit cells. It can also transfer the mutated section of DNA to other cells (not always of the same species) so that other cells can have the same advantage. The end result is "smarter bacteria" that don't care anymore that you're taking the antibiotic. Do this*

over years and years of people taking antibiotics for every little sniffle and all you've got left are the smart bacteria that are resistant to a lot of antibiotics.

Bacteria as Social Creatures

Bacteria and archaea are both prokaryotes that are *unicellular,* but that doesn't mean they aren't social. There are some communities that can be encased into a stable "slime" matrix called *biofilms.* All of the cells of a biofilm are of the same species, but they will show some evidence of differentiation so that some cells will do one thing while other cells will do other things. This primitive differentiation is thought to be due to signaling between the cells.

The bacteria in a biofilm will have an interesting process going on called *quorum sensing.* With this type of cell signaling process, bacteria sense the environment and can tell if the timing is right to start growing rapidly as a group or if they should hang back, waiting for more optimal circumstances. The cells can talk to one another so that the interior cells in the colony will wait for the right signal to divide and grow.

There are even situations where channels of liquid and nutrients will flow throughout the biofilm, acting like a primitive circulatory system. The communication allows the biofilm to have collective behaviors, including dying together as a unit, making some biologists to consider biofilms as a sort of multicellular organism in a primitive way. They can allow clusters of bacteria to resist antibiotics and other harsh environments, making them more dangerous should they cause an infection than cells acting by themselves.

Evolutionary Transition from Prokaryotic to Eukaryotic Life Forms

Probably the best way to see the difference between prokaryotes and eukaryotes is to look at how they came to be. Prokaryotes started to develop on Earth after the planet was just 750 million years old. This was about 3.8 billion years ago; how it happened for the first time is still a bit mysterious. We can't make life in a test tube by putting the right molecules in the same place, so it just isn't clear how it happened in nature.

Long ago, Earth had an atmosphere that came from volcanoes and deep space — a lot of carbon monoxide, hydrogen gas, hydrogen sulfide, nitrogen gas, and carbon dioxide. Early organisms used carbon under low-oxygen conditions to make energy in ways that are far less efficient than cells make energy today.

Eventually, events like sunshine and lightning strikes could have sparked enough energy to make larger organic molecules. When some of these molecules had the necessary chemistry to self-replicate, life was well on its way to becoming a reality. Among the different macromolecules, only nucleic acids have this ability.

The best guess biologists have come up with lately is the idea that the first life forms did not have DNA as their main nucleic acid; instead, they used RNA, which is perfectly capable of doing then what DNA does nowadays. This period of time was called the "RNA world." RNA was able to use a unique code, called the genetic code, to make proteins. The genetic code our cells use now is probably not much different than it was back then. They call this phenomenon the *conservation* of the genetic code over time. An early RNA-bearing cell likely looked like this:

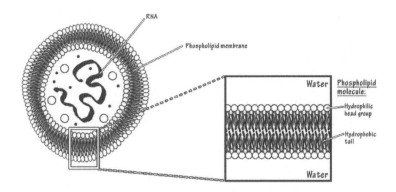

Another thing that was highly conserved over time was the ATP molecule. All cells we know of use ATP as their major energy currency. This energy was first made by cells using glycolysis only. Then, there were cells that used photosynthesis in order to make energy out of CO_2 (which was plentiful) and sunshine. Lastly came organisms that used the waste product of photosynthesis (which is oxygen) to make energy through cellular respiration, or what's called *oxidative phosphorylation*. Can you see how this changed the atmosphere and allowed the whole planet to evolve into what it is today?

As you will learn, eukaryotes are more complex. They have internal organelles, which are membrane-bound structures not seen in prokaryotes. These first came to Earth about 2.7 billion years ago. Archaea are actually more similar to eukaryotes than they are to bacteria, so it's probable that there was some common ancestor leading to both eukaryotes and archaea.

There is a theory called *endosymbiosis*, which says that small prokaryotes in primordial times were sometimes engulfed by larger prokaryotes. This led to the probability that some of these interesting cells developed symbiosis with one another so they couldn't live apart. As time went on, the cells evolved to have these internal structures become organelles instead of individual organisms.

Both chloroplasts and mitochondria were probably once freely living structures that got engulfed and incorporated into a eukaryotic cell. How do we know this probably happened? Both of these structures have their own DNA outside of the regular cellular/nuclear DNA. Both have ribosomes more similar to bacteria than to eukaryotic ribosomes. They also have double membranes. The inner membrane is similar to bacterial membranes while the outer one is similar to eukaryotic membranes. They can't live by themselves, though, because they have de-evolved. They don't have all the DNA inside them to make a whole freely living organism.

The first eukaryotes were single-celled organisms. Those that exist today include yeasts, amoeba, euglena, algae, and giardia. The algae are photosynthetic single-celled organisms while organisms like the amoeba are so specialized that they have unique ways of moving that are far different from any prokaryotic cells.

The first multicellular organisms probably came about 1.7 billion years ago. These likely started out as aggregates of single-celled organisms that evolved over time to become so differentiated and specialized that they needed each other to survive. Ancient algae probably did this, eventually becoming primitive plants that evolved over time to become increasingly complex and less like blobs.

To Sum Things Up

Prokaryotic cells are simple cells that don't have any internal enclosed membranous structures called *organelles.* They are also all single-celled organisms that may or may not have the ability to move independently. As we discussed, these organisms can be very social, forming colonies called *biofilms* that are able to act as a unit because they can communicate with one another.

The eukaryotes came after the prokaryotes on Earth. These more complex cells evolved after smaller prokaryotes were engulfed by larger ones. These internalized structures lost the ability to live outside freely, and the larger cells became so efficient at getting energy because of these internalized prokaryotes that they couldn't live without them either. The end result was symbiosis that led to the evolution of many more complex forms of life.

CHAPTER 14:
THE STRUCTURE OF A EUKARYOTIC CELL

Eukaryotic cells are so complex that it will take more than one chapter to cover everything that goes on inside them. This means that in this particular chapter, we'll just get introduced to eukaryotic cells so you know what they look like and what their different parts are. You will learn a bit about what these cells do as "tiny factories" and will see how all of the different structures interact or work together in order to allow the entire cell to function as a unit.

Just the Basics: What Eukaryotic Cells Look Like

Eukaryotic cells are interesting and beautiful examples of biochemistry in motion. In the different parts of eukaryotic cells, there are structures that make lipids, while nearby parts of the cell make proteins instead. Still other parts act as "post offices" — sorting out the different molecules that get made by the various organelles so they can get packaged into tiny vesicles to be sent to the far reaches of the cell, or even outside the cell itself.

Let's first look at this beautiful but compact cellular package:

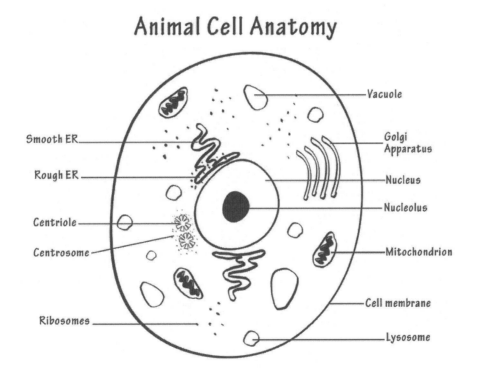

Here is a brief summary of these structures as they exist in almost all animal cells:

- **Plasma Membrane** — The plasma membrane is often called the *cell membrane*. It is mostly made from *phospholipids* but also has other lipids, proteins, glycoproteins, and cholesterol. These cell membranes are called *semi-permeable* because they easily allow some molecules to cross over, forcing other molecules to either never make it across or to have special ways of doing this.

- **Cytoplasm** — This is gel-like as it is in prokaryotes. In eukaryotes, it contains ions, proteins, carbohydrates, and anything else that can dissolve in a watery environment. It isn't a bag of liquid, though, because there is a skeleton of proteins called the *cytoskeleton* that will help compartmentalize the cell in order to hold all of the organelles in their rightful spaces.

- **Cytoskeleton** — These are amazing structures made from different proteins. Like all proteins, they are long polymers made from smaller subunits in order to make an interwoven mat of these skeletal structures that allow for many different cell processes to take place.

- **Endoplasmic reticulum** — These are a series of membranes located in clusters near the nucleus. They can be studded with ribosomes, so they look like membranes with beads along them. This endoplasmic reticulum is called *rough endoplasmic reticulum* because it looks rough in appearance. On the other hand, there are stacks of these membranes called *smooth endoplasmic reticulum.* These areas do have not ribosomes and make lipids as their main job. The rough endoplasmic reticulum does more than hold onto ribosomes. This is where proteins first go after being made in order to be further processed before being sent off into the cell.

- **Golgi complex** — A Golgi complex can also be called a *Golgi body*. This is a really interesting structure often called the "post office" of the cell. Molecules enter one side of it and get processed through a series of stacked membranes called *cisternae.* They exit the other end of the apparatus completely packaged into *vesicles* (small lipid chambers) that travel to different places inside and outside the cell.

- **Centrioles** — These are tiny structures shaped like cylinders. Their job is to make one of the main types of the cytoskeleton called *microtubules.* They become important during cell division because they make the microtubules needed to separate the DNA chromosomes when each cell divides.

- **Cilia and flagella** — These sound similar to the same structures in prokaryotes and do essentially some of the same things but they are structured differently and use unique ways of moving that is not at all the same as what happens in prokaryotic cells.

- **Lysosomes** — These are lipid sacs that contain very strong enzymes. These enzymes will chew up and digest larger macromolecules in the cell when they are no longer needed in order to recycle the parts. They also sometimes take in parts of bacteria and other infectious organisms in order to chew these organisms up as part of a specialized immune response in multicellular organisms that have an immune system.

- **Mitochondria** — You now know the reason why these are called the "powerhouses of the cell." This is because this is where most of the energy in the cell comes from. Cellular respiration processes and the Krebs cycle both happen inside mitochondria, which use oxygen rapidly in order to create carbon dioxide, water, and lots of energy.

- **Nucleus** — These are the membrane-bound structures that house the DNA of eukaryotic cells. They have a special outer membrane called the *nuclear envelope.* In eukaryotes, the DNA isn't in one big circle but is separated out into many different linear chromosomes that are balled up in a big wad of DNA until the chromosomes get copied and separate when the cell divides.

- **Nucleolus** — This is a smaller structure within the nucleus. For a long time, biologists didn't know what they did but now they know this is where ribosomal RNA is made and where ribosomes themselves start being assembled for the purposes of making protein.
- **Ribosomes** — These are tiny protein-making factories that take the messenger RNA "message" gotten directly from the nuclear DNA and turn this message into proteins. This is basically all they do; they just continually churn out polypeptide chains as long as there are amino acids available and messenger RNA that needs to be translated into protein molecules.
- **Peroxisomes** — These look a lot like lysosomes and they also have a lot of enzymes in them that destroy larger molecules. The enzymes in these cells are oxidizing enzymes that make highly oxygenated byproducts like hydrogen peroxide, or H_2O_2, as a byproduct. Like lysosomes, they must be separated from the rest of the cell so that their enzymes don't just digest the entire cell from the inside out.

What about plant cells? These are very similar to animal cells but have some unique structures inside them. This is what a typical plant cell looks like:

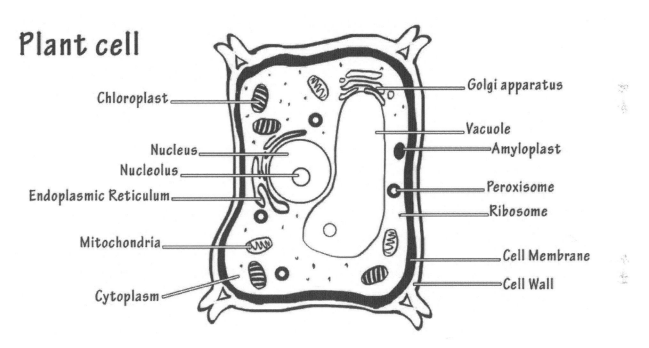

Inside the plant cell, these are the structures you'll see that you won't find inside an animal cell:

- **Cell wall** — While plants have a cell wall to keep them from bursting or shrinking under high or low water conditions, these are different in structure from those seen in archaea and bacteria. Instead of peptidoglycan, you'll see cell walls made of *cellulose* in most plant cells. In fungi, you'll see *chitin* as the main cell wall structure. Here is what the cell walls of many plant cells look like:

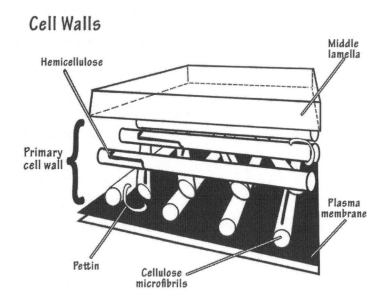

- **Vacuoles** — These are large bags of mostly water. They act as reservoirs to take in water when there is too much of it and donate water when there is a relative drought. They help to make the plant cell more balanced with respect to water so they can better adapt to their environment, which can be harsher than you'll see around many animal cells.
- **Chloroplasts** — These are interesting structures inside plant cells and most other eukaryotic cell types that participate in *photosynthesis.* This is where these photosynthetic biochemical processes happen to take carbon dioxide from the atmosphere in order to make mostly starch polymers of glucose to be used later as fuel sources. This is what a typical chloroplast looks like:

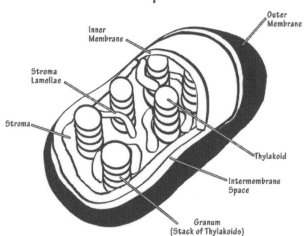

In the next section, we'll dig deeper into these cell structures. You'll see how these different cell parts not only do their job but will interact with one another in order to have a symphony of different actions that together do all of the work each individual cell must in order to survive.

Getting Down with the Details of Eukaryotic Cell Structures

In this section, we will talk about the different structures in more detail as you start to piece together how these interact with one another. We will start with the plasma membrane.

Plasma Membrane

The plasma membrane, or cell membrane, is one of the most important structures of the cell as it is the main barrier between the cell and the potentially hostile environment around it.

As you know, the plasma membrane is mostly lipids in a bilayer that allows the parts of the phospholipid that is hydrophilic to be in contact with the aqueous inside and outside of the cell while keeping the inner hydrophobic layer isolated from these areas. In animal cells, cholesterol helps make the cell less fluid. The proteins associated with the membrane can sit within the membrane; these are called *transmembrane proteins.* There are peripheral proteins that sit on the surface of the cell membrane, usually on the outside of the cell. This is what the typical cell membrane looks like:

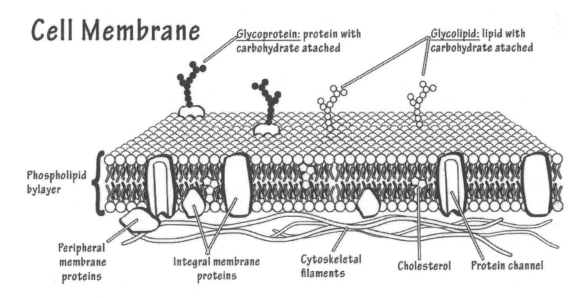

Notice how some of the external proteins sticking up outside of the cell membrane have what looks like tree branches sticking out of them? These are chains of sugars, which turn the protein into what's called a *glycoprotein* (in order to reflect the sugar component of the protein). These glycoproteins are very important to cell signaling and help to make receptors on the outer part of the membrane that will bind to things like hormones and other signaling molecules.

Phospholipids are not all created equal. There are four types of phospholipids in animal cells, called *sphingomyelin, phosphatidylethanolamine, phosphatidyl serine*, and *phosphatidylcholine.* All together, these phospholipids account for more than half of all the molecules making up the cell membrane. One additional type is called *phosphatidylinositol.* This is negatively charged and sits on the cell interior in order to help the inside of the cell to be more electrically negative on the inside compared to the outside.

The cell membrane also makes use of combination molecules made from lipids and sugars, called *glycolipids.* Like the glycoprotein molecules we've talked about, these probably help cells pass signals from one to another by sitting on top of the cell membrane (facing outward). Only about two percent of all lipids in the cell membrane are made from these lipid types.

Plasma membranes have a key feature you should know about: its semi-permeable characteristics. This means that some molecules will easily pass through the membrane, while other molecules need a great deal of help to do this. This is what's meant by the term *semipermeable*:

Cell Membrane Function

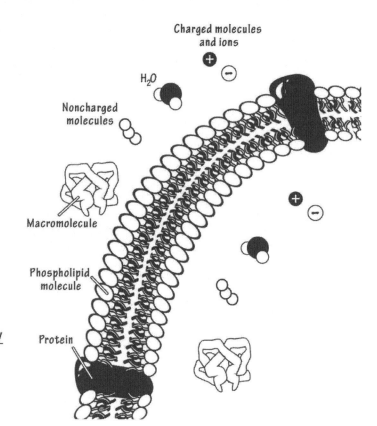

Impermeable - nothing can pass through

Permeable - most things can pass through

Semi- permeable - smaller molecules pass through but not larger molecules

Selectively permeable - only certain small molecules and certain large molecules can pass through.

Cell membranes are selectively permeable

Cholesterol floats within the membrane and has some unique properties. It will block the ability of the membrane to be fluid under high temperature conditions so that the permeability will be reduced. Under low temperatures, the opposite happens, so having cholesterol in an animal cell membrane will block the cell from freezing.

The proteins in the cell membrane have numerous functions for the cell. These include providing proteins that label the cell on the outside; proteins that act as channels for ions, water, and other molecules to get through the membrane; and proteins that act as either receptors or signaling molecules. Proteins that act in the cell signaling process will take a signal from the outside of the cell and transfer it to a signal or cascade of signals that happen inside the cell.

There are three kinds of proteins associated with membranes. Proteins called *extrinsic proteins* sit outside the membrane. These tend to be polar proteins that just don't mix with the nonpolar environment of the cell membrane (although they have parts necessary to stick to the phospholipid bilayer).

Integral proteins are somewhat inside the membrane itself because they are partially nonpolar and will dissolve or integrate themselves with the cell membrane. Those that are called *transmembrane proteins* will cross completely through the entire membrane. These often act as channels to allow molecules that are polar to get through the cell membrane. This is what these look like:

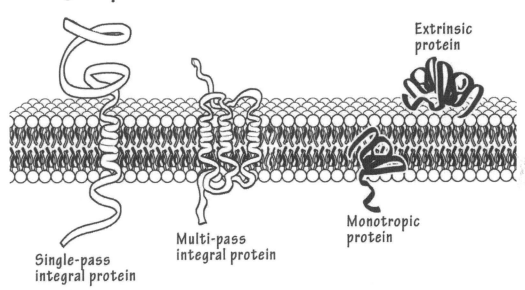

One interesting function of the cell membrane is to maintain the water pressure inside and outside the cell. This is called *maintaining tonicity.* You know all about how this works if you do any gardening. Watch what happens if you don't water your plants for a period of time. Notice how dead and wilted they look?

Now, add a bunch of water to your wilted plants and wait a few hours. See how they plump up and act as though they've been revived from the dead? All of the wilting happened because the cells of the plant were seriously dehydrated. Adding water causes these water molecules to travel through the process of *osmosis* from the outside of the cells to the inside of the cells. The end result is a plumper, healthier cell.

Water movement across a semipermeable membrane is called *osmosis.* Osmosis is like diffusion, but it only applies to water and water across membranes. Water will travel from a very dilute solution (with water as the solvent) to a very concentrated solution.

This sounds like the opposite of what should happen, but if you think about it, a concentrated solution of a salt, for example, has fewer water molecules in it than a dilute salt solution. This means that water molecules will go from where there are a lot of water molecules to one with fewer of them. This is what it looks like:

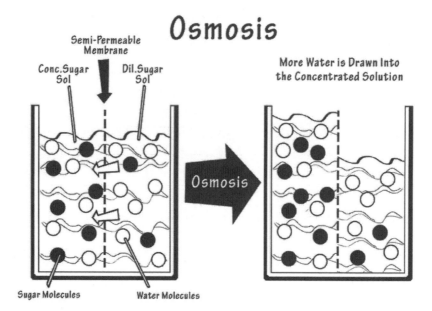

A related concept to osmosis is the idea of *tonicity*. This applies easily to cells in the human body or in any kind of salty solution. If the concentration of particles in a solution outside the cell is the same as that which Is inside the cell, the outside solution is said to be *isotonic* and the cell will stay the same size.

If you add these same cells to a dilute solution, this dilute solution is said to be *hypotonic.* The cell will burst because water will rush into it through osmosis.

If the solution outside the cell is concentrated compared to the concentration of particles inside the cell, it is called a *hypertonic solution* and the cell will shrink. This is what it might look like:

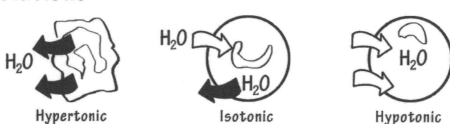

The Cell Nucleus and Nucleolus

The *cell nucleus* is an organelle that mainly houses the genetic material, or DNA, of the cell. All eukaryotic cells have this organelle. Besides having the DNA housed there, this is where the genetic message gets transcribed into the messenger RNA molecule. After this happens, the messenger RNA travels outside the nucleus through pores called *nucleopores*. Small molecules and ions also pass through these nucleopores. The nuclear membrane is also called the *nuclear envelope.*

A process called *DNA replication* also happens in the nucleus. DNA replication is the same as DNA copying. The only time it needs to do this is when the cell is about to divide. There are enzymes in the nuclear milieu (called *nucleoplasm*) that attach to the DNA chromosomes, split the two strands apart, and add new matching strands to each split strand in order to make two identical copies of each chromosome.

A unique structure inside the nucleus is called the *nucleolus*. This isn't membrane-bound, so it really isn't technically an organelle. It is simply a space where the ribosomal RNA is made. Ribosomal RNA is very big and makes up the majority of the cell's ribosome. These ribosomes get assembled inside the nucleolus before leaving altogether in order to make proteins.

It's this ribosomal RNA that becomes an enzyme even though it isn't made of protein. Enzymes made out of RNA instead of DNA are called *ribozymes*. This is what these ribosomes look like inside the nucleolus:

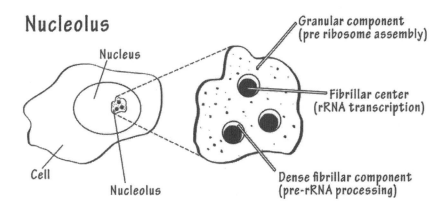

The Endoplasmic Reticulum

As you know, there are two kinds of *endoplasmic reticulum*. The RER, or *rough endoplasmic reticulum*, is where proteins are made and processed. They are closely affiliated with ribosomes. Once a protein chain gets made, it enters the RER in order to undergo a process called *post-translational modification of proteins.* We will talk more about how this works later.

The *SER*, or *smooth endoplasmic reticulum*, is smooth because it doesn't have ribosomes dotting its surface. It's in these membranes that lipids of all types get made and processed. Both SER and RER are huddled around the nucleus so they can be located near this structure.

This is a closeup view of these interesting organelles:

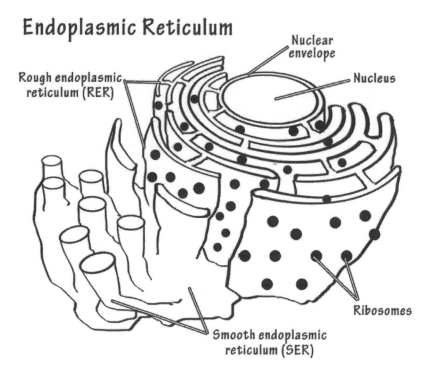

In some cells, such as the muscle cells of your body, the endoplasmic reticulum is called the *sarcoplasmic reticulum* instead. These membranes house calcium ions. When the muscle cell gets a signal to contract, the sarcoplasmic reticulum allows calcium to flow out of it and help to contract the muscle cell. These same calcium ions get pumped back into the sarcoplasmic reticulum in order to reset the muscle cell so it can contract again.

Ribosomes

The major role of ribosomes is to translate the messenger RNA message coming from the nucleus into protein or polypeptide chains. These are true factories that bring together the messenger RNA, the ribosomal machinery, and amino acids. Like a telegraph office, the ribosomes take the message encoded in the messenger RNA molecule and, using a secret code called the *genetic code*, they make proteins.

Once the protein chains are made, they get funneled into the RER. It is here that the proteins are checked for accuracy. If they are damaged or dysfunctional in some way, the protein is sent back to be recycled and is essentially rejected by the RER. Those that need modifying get modified by adding things like methyl groups or phosphate groups to them. They can either be released directly or can be sent onward to the Golgi apparatus, where they get packaged and shipped out.

This is how that process happens inside the endoplasmic reticulum:

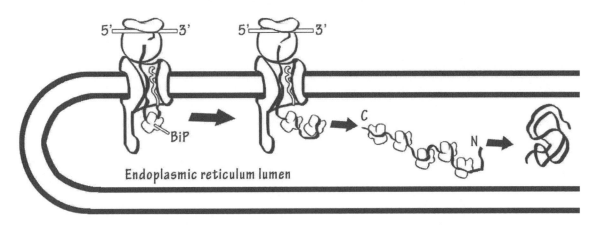

Protein Modification in the Endoplasmic Reticulum

Golgi Apparatus

The *Golgi apparatus*, or *Golgi body*, is an amazing organelle. It is a small structure located near the endoplasmic reticulum. There are about 10 to 20 of these Golgi bodies per cell. These are flat sacs of membranes responsible for sorting and packaging the different molecules made by the cell. It's a lot like a cellular "post office" that decides which molecules go together and where they go. This is what it looks like:

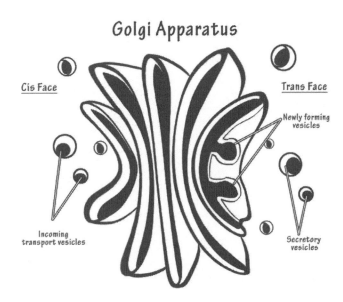

Golgi Apparatus

You can see that there is a *cis face*; it faces the endoplasmic reticulum and is where the molecules first enter the Golgi body. They traverse through the Golgi body from the cis face to the opposite side, called the *trans face*. The trans face is where these packaged vesicles leave the Golgi body for good. There are systems in place that traffic these vesicles throughout the cell or send them to the cell membrane, where the contents of the vesicles get discharged into the cell exterior completely. Exactly how they are tagged so that the cell knows where to send each vesicle isn't completely known.

Once a protein or other molecule enters the Golgi apparatus, it could leave immediately if it was sent there by mistake. The Golgi body just ejects it. Others are destined to leave the cell, so they get packaged into tiny vesicles that go to the cell membrane. The vesicle will merge with the membrane and discharge its contents in a process known as exocytosis.

A few proteins stick to the vesicle wall, which is a membrane, too. As the vesicle gets discharged at the cell membrane (in *exocytosis*), these proteins remain in the *vesicle membrane*, which has now become part of the cell membrane. A lot of these proteins become *transmembrane proteins* like channel proteins — put to work immediately in the cell membrane. Still, other proteins and other molecules go directly to the lysosomes for destruction.

Lysosomes

Lysosomes are a lot like big vesicles. They honestly don't look like much. Even so, they are very powerful organelles that have strong enzymes inside them. These enzymes catalyze or break down other molecules — large and small ones — in order to turn them into small breakdown end products. The pH inside lysosomes is low (about a pH of 5). This is often acidic enough to participate in breaking down some of the larger molecules by themselves without any enzymes necessary.

Fun Factoid: There are about 50 different diseases in humans called "lysosomal storage diseases." These are diseases where one or more of the enzymes inside the lysosome is missing or nonfunctional. The end result is that the molecule the deficient enzyme is supposed to break up doesn't ever get degraded; it builds up in huge quantities inside the lysosomes, leading to many severe health issues. About one in 8,000 babies are born in the world each day with one of these storage diseases.

Peroxisomes

Peroxisomes are similar in appearance to lysosomes but have a slightly different job. These tiny vesicles have powerful oxidizing enzymes in them that mainly chop up fatty acids and amino acids into smaller pieces. As oxidizing enzymes, their function leads to reactive oxygen species like *hydrogen peroxide* as an end product. H_2O_2, or *peroxide,* is what gives these structures their name. Molecules like this would be very damaging to the cell if they weren't kept isolated within the confines of these small organelles.

Fun Factoid: Peroxisomes also break down a lot of toxins in your body. After a night of binge-drinking, for example, your liver peroxisomes are very busy breaking up all the ethanol you drank. This generates a lot of free-radicals, such as O_2- ions and H_2O_2 molecules. These would be too destructive to stay that way, so your peroxisomes also have antioxidant molecules that break the oxygen free radicals into water, which is much safer for your cells.

Mitochondria

You know already that mitochondria are the "powerhouse" organelles of the cell because they are tiny energy-making factories. Red blood cells in your body have none of these organelles, while busy and active muscle cells have thousands of mitochondria. There are as many mitochondria in a cell as are needed for cellular energy.

It's in these mitochondria that oxygen gets taken up in huge amounts. These oxygen molecules get processed as part of cellular respiration in order to break down carbon-based molecules into carbon dioxide and water. There are many mitochondrial diseases in humans where these organelles don't function well. As you can imagine, muscle weakness is a major symptom of these disorders because muscle cells will not have the energy they need to function to their greatest potential. This is what a mitochondrion looks like up close:

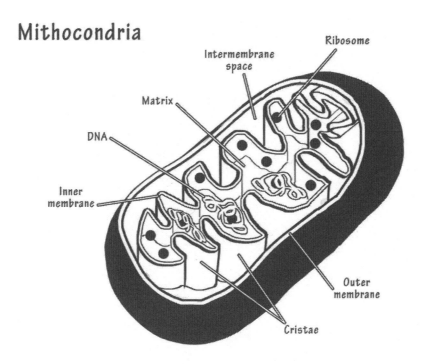

Cytoskeleton of the Cell

Cytoskeletons are interesting structures that come in three basic categories (and a lot of subcategories). There are very thin microfilaments made of proteins like *actin* and *myosin.* They help to control movement of substances inside the cell and help muscles to contract in muscle cells. They are too flexible to play a big structural role in the cell. There are other cytoskeleton types that do a better job of this. This is what the *microfilaments* look like:

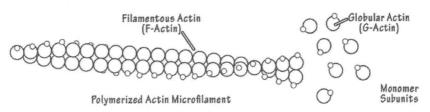

Intermediate filaments are thicker than microfilaments. They can be made from many different types of proteins, depending on what their job is. They serve to give the cell its structure and often attach to the cell membrane in order to help anchor the cell to a neighboring cell in interesting ways. This is what intermediate filaments look like:

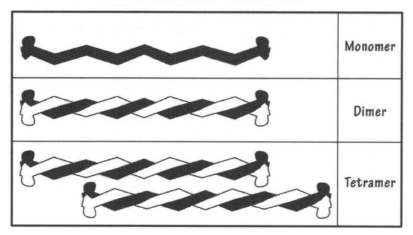

Microtubules are the thickest of all of the cytoskeletal elements. They help greatly as the drivers behind *mitosis*, where the chromosomes get separated during cell division. They also drive the cilia and flagella of the eukaryotic cells, where they are energized using ATP energy. There are two proteins that make up microtubules, called *alpha-tubulin* and *beta-tubulin*. This is what they look like:

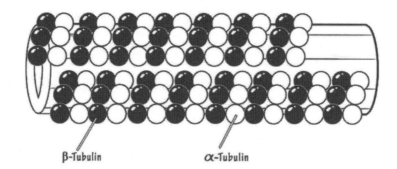

Chloroplasts

Chloroplasts are only found in eukaryotic cells that undergo photosynthesis. We will talk a great deal more about how photosynthesis works but it basically involves organelles that have pigment in them. The pigment gathers light energy from the sun in order to drive reactions that start with carbon dioxide and water. Driven by light energy, the chloroplasts make glucose molecules and give off oxygen gas as a byproduct. The glucose molecules gather together in a polymer called *starch*. We eat this starch every time we eat a vegetable. Potatoes are called "starchy vegetables" because they have so much in them. This is what chloroplasts look like:

Chloroplast

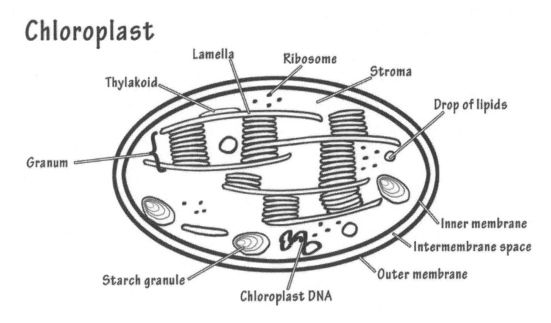

Transportation of Molecules in Eukaryotic Cells

We will talk next about the cell membrane and how important it is in the transportation of molecules in and out of the cell. Inside the cell, however, transport is not a random occurrence. Vesicles made and/or processed in the Golgi apparatuses need to get from place to place. They are often packaged in small transport packets that travel through the cell using actin microfilaments as guidewires. Often, microtubules are used as well to help these vesicles get from place to place. Prokaryotic cells don't have this intricate network of filaments; they are forced to use more limited means of intracellular transport.

Sometimes, the cell needs to have molecules taken in from outside. The two main processes that do this are called *endocytosis* and *pinocytosis*. Endocytosis involves grabbing a collection of cell membranes and performing the reverse of *exocytosis* — essentially drawing in mouthfuls of molecules in one big gulp. These get transported into the cell as vesicles or can just get discharged after being taken up during the endocytosis process.

The processes of pinocytosis, or "cell drinking," is similar to endocytosis but on a smaller scale. It allows smaller molecules to enter the cell along with a sip of water that gets incorporated into the cytoplasm.

Regardless of how these molecules get in, they often make use of the cytoskeleton to figure out where these newly taken-up molecules finally end up inside the cell. How the cell knows where to send which things isn't clear. It could have to do with receptors on the vesicle that label them according to their final address. Alternatively, it could relate to other types of chemical signaling inside the cell.

Cell Compartmentalization

Because of the intricate network of filaments inside eukaryotic cells, they aren't simply "bags of liquid" with stuff in them. They are highly organized. You wouldn't want the endoplasmic reticulum to be on the far side of the cell from the nucleus, would you? It would mean that the messenger RNA from the nucleus would have a long and treacherous journey before getting to the ribosomes. It makes much more sense to have these near to one another.

Part of the solution is to trap the different organelles into the network of filaments within the cell. These filaments help keep each organelle where it belongs. If something needs to be moved, the microtubules are connected to a motor-like protein called dynein that acts like a pulley system, dragging even large organelles from one place to another.

To Sum Things Up

Eukaryotic cells are far more complex than prokaryotic cells. Their organelles are like different sections of a larger factory. Each has a job to do and many will send molecules from one place to another as they go from a DNA message to an RNA "messenger" and onto polypeptides and fully functional proteins.

There are also structures in the eukaryotic cell that act to break down molecules and substances that are no longer needed or are damaged in some way. Both *lysosomes* and *peroxisomes* are vesicles of a sort that contain many enzymes used to break up large molecules into smaller molecules and atoms. Some of these get recycled at a later time.

The *cytoskeleton* gives the cell structure. Some of them help anchor two cells to one another, while others help in allowing the cell to move independently. *Microtubules* also act to pull chromosomes apart in cell division and serve to allow vesicles to travel within the cell to wherever they are needed.

SECTION FIVE:
CELL TRANSPORT AND COMMUNICATION

While you might think you know all about the plasma membrane of the cell and how it works, these next few chapters might expand your understanding more than you'd expect. The plasma membrane is the mouth, the GI tract, and the urinary tract (all at once) of the cell because it is responsible for everything that enters and leaves it. It needs to be as picky about how it does this as you are choosy about what you eat (hopefully).

CHAPTER 15:
THE PLASMA MEMBRANE:
THE GATEKEEPER OF THE CELL

Rather than think of the plasma membrane as a wall, think of it instead as your cells' gatekeepers. They are not solid barriers but allow substances to pass through them according to the needs of each cell. Passage of materials goes both ways; ideally, nutrition goes in and waste products go out, while electrolytes will pass either way.

The plasma membrane goes by many names, so don't be confused. You can call it a *cell membrane, plasma membrane, cytoplasmic membrane,* or *plasmalemma.* Its main job is to form the separation boundary between the inside and the outside of the cell. Most of it is a lipid bilayer but there is a lot of matter floating in it. The membrane is fluid. This is what biologists call the *fluid mosaic model.* It means that most of what is embedded in the membrane literally floats in a sea of *phospholipids.*

What is the Plasma Membrane Made Of?

About 40-50 percent of the total amount of the weight of the plasma membrane is lipid based (most of it being the phospholipid bilayer). About 40 percent of these lipids is *cholesterol* in animal cells. This is important because, if you really try to restrict your cholesterol levels, you might not be helping your cells. The rest of the plasma membrane consists of proteins (about 50-60 percent by weight).

When you look at a plasma membrane, though, you will see mostly lipids. This is because the above values are by weight and not by molecules. Proteins weigh more than lipids and are denser, so there are fewer numbers of these molecules. They just weigh more than the lipids.

The phospholipids in the cell membrane are not generic. There are different kinds, called *phosphatidylinositol, phosphatidylethanolamine,* and *phosphatidylcholine.* There are also molecules called *sphingomyelin* and *glycolipids,* which are lipids attached to sugar molecules. This is what these look like:

Phospholipids

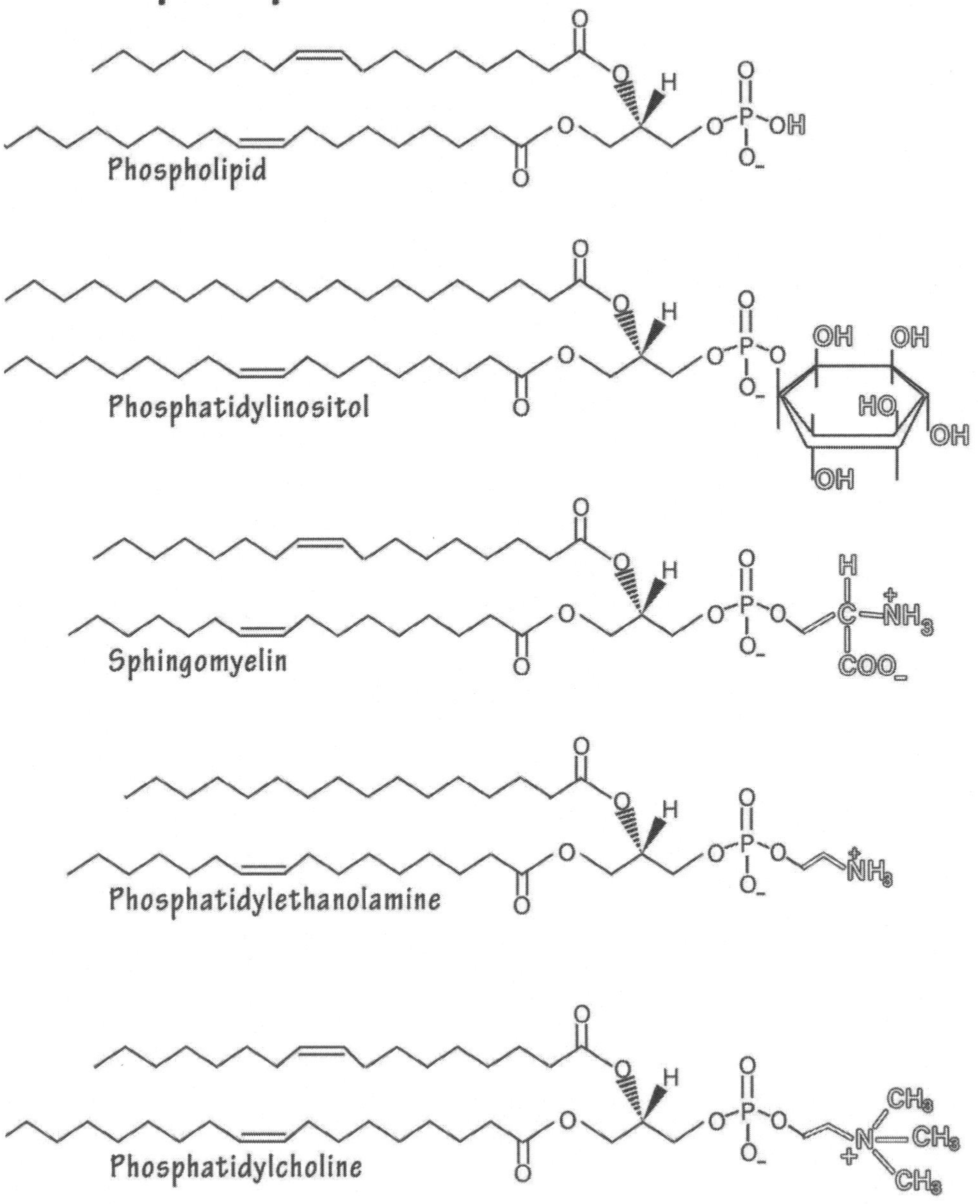

Phospholipid

Phosphatidylinositol

Sphingomyelin

Phosphatidylethanolamine

Phosphatidylcholine

You should know that not all plasma membranes are the same. Remember that saturated fats are solid at room temperature (think vegetable shortening or lard) and that unsaturated fats are liquid at room temperature (think olive oil or canola oil). The more saturated fat in a cell membrane, the less fluid it will be (and vice versa). Shorter chain phospholipids make the cell membrane more fluid overall.

Cholesterol is interesting. It makes the cell membrane less fluid at high temperatures (like body temperature) but more fluid at low temperatures. So, in animals that might otherwise freeze, cholesterol keeps their cells from freezing. In a sense, it acts like antifreeze in the cell. Cholesterol can float in clusters along with proteins that are collectively called *lipid rafts* in the membrane.

This is what a lipid raft of cholesterol looks like:

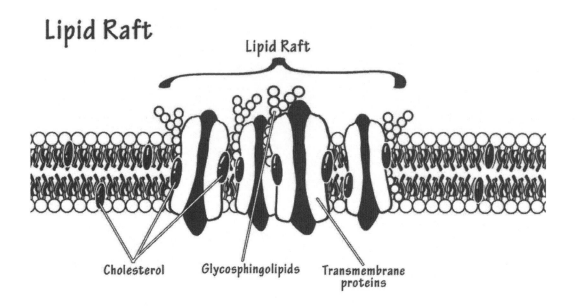

How Does the Cell Membrane Get Made?

There are no enzymes busy making cell membranes for you every day. They simply assemble themselves. How do they do this? It's a natural part of chemistry. If you take a bunch of phospholipids and put them in water, they will naturally line up because there are parts that like water and parts that don't. This is exactly the reason why you'd see balls of them like the beads of oil you see in your vinaigrette, which is just a mixture of oil and water anyway. There are *van der Waals forces* and *hydrogen bonding* that keep them lined up next to each other.

The simplest form you would see in a watery solution is called a *micelle.* This is a simple ball of lipid with the *hydrophobic* ends inward and the *hydrophilic* ends outward. This is what it looks like:

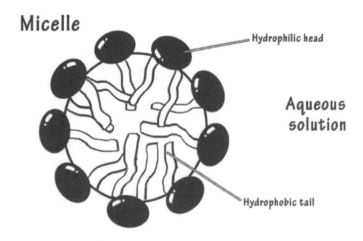

If you make the layer of phospholipids into an actual "simple cell," you would get an inside and outside compartment — each of which would be watery — plus the lipid bilayer.

This is what it looks like:

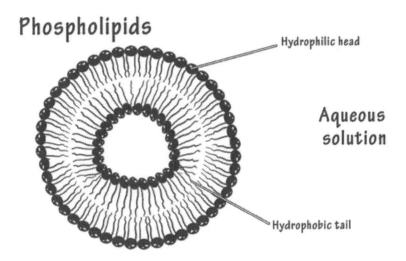

The cell membrane gets more complex as *proteins, cholesterol, glycoproteins,* and *lipoproteins* get added. Some are added actively through cell processes, while others self-assemble into the membrane. Regardless of how this is done, the hydrophobic parts stick together, and the hydrophilic parts do the same thing. The end result is a more complex structure that looks somewhat like this:

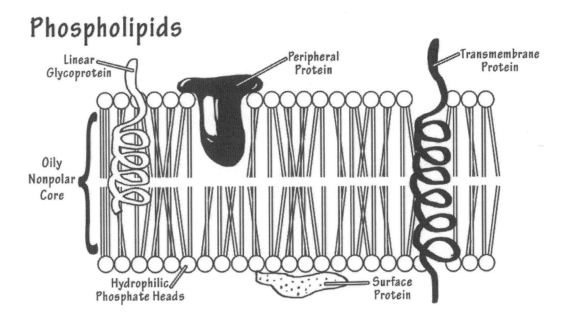

Cell Membrane Carbohydrates

There are carbohydrates in the cell membrane. Most of these are attached to proteins called *glycoproteins,* although there are some glycolipids as well. These glycolipids have names like cerebrosides and *gangliosides. Glycolipids* like this are found in nerve cells and immune cells, primarily, and seem to be very important in the way they function.

Some cells have a thick *glycocalyx* layer. This is a layer of glycoproteins on the cell surface. If you see a lot of these, you know the cell needs this layer for adhesion of things to the membrane. You can think of the glycocalyx as a fuzzy layer on the cell surface because the molecules of sugar stick out of the membrane surface. In white blood cells and platelets in humans, the glycocalyx helps these cells stick to one another and to other things.

This is what the glycocalyx looks like. See how fuzzy it is?

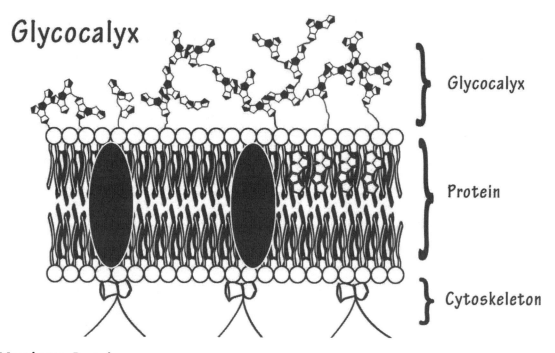

Cell Membrane Proteins

There are a lot of important proteins in the cell membrane — so important that nothing much would happen in the cell if it weren't for these proteins and the many jobs they do. Here's a summary of these protein types:

- **Integral proteins** — These are also called *transmembrane proteins* because they span the cell membrane. There are usually hydrophilic parts that stick out of either end of the lipid bilayer. They make up ion channels, certain receptors on the cell membrane, and the different ion pumps that help to pump ions into or out of the cell.
- **Lipid-anchored proteins** — These are attached to the lipid bilayer or other lipids in the membrane through *covalent bonding.* There is a lipid part of this protein that does this; the protein itself does not make contact with the membrane itself.
- **Peripheral proteins** — These are proteins that are loosely stuck onto the membrane using bonding types that are easier to break up (think hydrogen bonds). Some of these temporarily attached proteins are enzymes and some hormones.

Proteins do a lot for the cell and the cell membrane. Proteins facilitate the movement of water, ions, and larger molecules through the cell membrane in regulated ways. They are also important in all aspects of cell-to-cell signaling. It takes proteins or glycoproteins to make receptors for *ligands* that bind to them. A ligand is anything that binds to a receptor in a specific way.

It also takes proteins (usually transmembrane proteins) to pass the signal activated on the outside of the cell onto the interior of the cell where the signal changes the cell and its function in some way. It's a lot like "telephone tag," where the message starts out on the outside, gets passed through the cell membrane, and activates an enzyme or biochemical pathway on the cell's inside in order to change the cell's behavior.

Most proteins do not just migrate to the cell membrane in order to stick themselves in there. If you think about how proteins are made, you'll understand how this might work. Proteins are made in ribosomes and go into the endoplasmic reticulum, which is just a bunch of membranes. Some of them stay in these membranes, get packaged by the Golgi apparatus while still stuck in a membrane, and leave the Golgi apparatus (still in their membranes). When the vesicles fuse with the cell membrane, the proteins are now part of it as well. See how it works in this image:

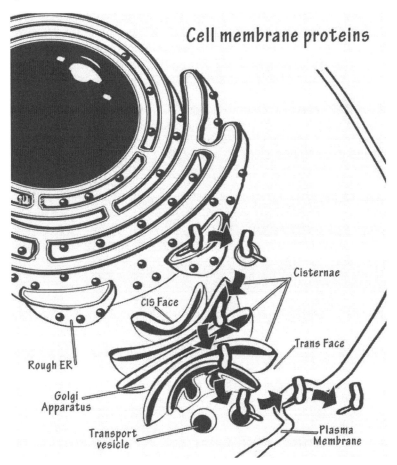

Cell membrane proteins

What Does the Cell Membrane Do?

By now you know that the cell membrane defines the inside and outside of each cell. It has anchoring points inside and outside the membrane for connecting the cells to other cells and for attaching the cytoskeleton of the cell at various points. (What good would the cytoskeleton be if it couldn't attach at several point to the membrane itself?)

The membrane is picky about what it will allow to pass through it. This is why it is called *semi-permeable*. It gets this pickiness from the chemistry of the membrane itself. If you're a lipid-loving molecule and not too big, you have a good chance of getting through this lipid layer. If you're a water-loving molecule on the other hand, the chances are not so good, especially if you are a charged or very polar molecule. Gases like oxygen and carbon dioxide get a pass and cross the membrane easily.

The transmembrane proteins can be *channel proteins, carrier proteins,* or *proteins* involved in cell signaling. Some of these proteins will work without the addition of energy. Others use ATP energy and are called "pumps" — like the *sodium-potassium ATPase pump*. The transmembrane proteins that are used in cell signaling are very creative in how they work. We will have an entire chapter on these amazing proteins.

The cell membrane will also take things into the cell and expel them in large numbers. The processes that do this are called *endocytosis* and *exocytosis*, respectively. These are not simple processes and involve special proteins and systems that decide which vesicles to suck into or out of the cell as well as where the process should happen. Endocytosis must happen with energy applied to it. This makes it an active transport process (remember that all active transport needs energy from ATP to drive the process).

To Sum Things Up

The cell membrane is the barrier that defines the inside and outside of each cell. They are made of a phospholipid bilayer "sea" throughout which proteins and lipids float in order to spread out across the cell membrane. These proteins have many functions that carry out activities that the lipids can't do themselves.

The next few chapters will talk about how things get into and out of the cell membrane. From what you already know, how this happens depends a lot on the chemistry of each molecule or atom trying to cross the cell membrane and on the proteins and energy available to let these molecules cross over. These are processes happening all the time in a regulated way, so the cell keeps the things it needs and gets rid of the things it doesn't.

CHAPTER 16:
DIFFUSION AND OSMOSIS

Diffusion and osmosis are related terms, but they are not the same. *Diffusion* involves the movement of any type of molecule or atom (or anything at all really) from one place to another down a concentration gradient. It's based on the idea that, given enough time and energy/heat, all things will spread out. Now that you understand thermodynamics, you can see why.

Let's say you had a vacuum space as big as a room with no molecules of anything in it. Add a jar full of pure oxygen gas and take off the lid. Realistically, would it make more sense that the oxygen gas molecules would hang out in the jar forever or would they leave the jar and spread out across the room? Obviously, they would spread out, but how and why?

If you took the same room and the same jar of oxygen and instead kept the entire system at 0 degrees Kelvin (absolute zero), what would happen? Remember: At absolute zero, molecules and atoms do not move. They have no kinetic energy and shut down completely. This oxygen gas wouldn't actually be a gas at all; it would be solidified. This is because gas molecules must have the energy it takes to move around and bump against one another.

According to Gay-Lussac's Law, pressure and temperature of a gas are proportionate to one another if you keep the volume the same. No temperature means no pressure either. At zero degrees Kelvin, there would be no pressure, the oxygen would be stuck in the jar as a solid—and, quite frankly, the jar would have to be super strong not to implode under the extreme vacuum inside the jar. In the theoretical "vacuum room," though, the pressure inside the jar and outside would be the same, so no worries there.

This is what diffusion looks like. Think it of as "responsible social distancing" on a molecular level:

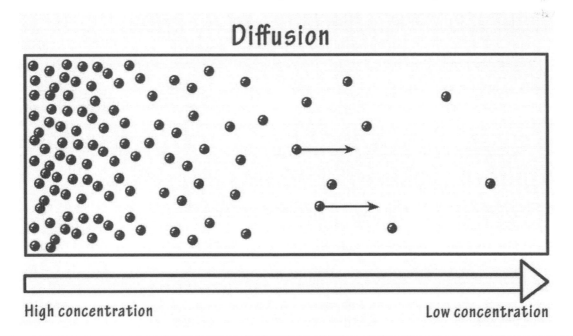

Diffusion

High concentration   Low concentration

With solid oxygen and no heat to give the molecules any energy to spread out, the jar would stay the same and no diffusion would happen. What you can conclude from this is that without some temperature, diffusion cannot happen at all. The hotter the system, the more energy these molecules have, and the faster they will diffuse over time. Temperature in general determines the rate of diffusion and not the end result. Given enough time and any heat at all, diffusion happens to the same degree every time as long as everything else is the same.

With temperature, everything depends on the second law of thermodynamics, which relates to entropy, or *disorder.* With molecules, temperature means they move, and movement means they bump into one another. As they do this, they spread out. As they spread out, the molecule density decreases, and they don't have a much of a chance of bumping into one another.

The whole process leads to a gradual spreading out of the molecules until they are roughly equally distant. This is when *entropy* is the greatest and, in the case of the oxygen in a jar experiment, the molecules will spread out, or *diffuse* until the whole room has an equal density of gas throughout. From this, we can conclude that diffusion goes from a more concentrated area to a more dilute area with respect to the molecules.

Diffusion happens in liquids as well and can also happen to a limited degree in solids. While diffusion isn't the reason some panes of window glass get more rippled over a long period of time (gravity is the reason), you can see that molecules in solids do move under certain circumstances.

The gradient also influences the rate of diffusion, but not the end result. If the jar of oxygen was placed in a room with a little bit of oxygen in it (and not in a vacuum), the molecules would still spread out, but they wouldn't spread as quickly. This is because the molecules of oxygen outside the jar will also bounce against one another and with the newly placed oxygen molecules. This slows the pace of the diffusion but won't prevent it from happening. From this, you can conclude that greater concentration gradients (differences) mean diffusion will happen faster.

Fun Factoid: The Greeks may or may not have understood diffusion as we know it today, but they did have a word for it. They called it "diffundere," which literally means "to spread out."

Diffusion happens on a small, local scale but can also happen over large differences. *Bulk flow* is another concept that involves movement of molecules in ways different from diffusion. When you breathe and gases are exchanged in the alveoli, both things happen. Gases diffuse between the capillaries in the lungs and the alveoli (air sacs), but when you inhale and exhale, that isn't diffusion at all. You use your diaphragm to allow for bulk flow of the gases into and out of your body. The same thing happens in your heart, which uses its pumping activity to allow for bulk flow to the entirety of your body. Imagine what would happen if we relied on diffusion alone to get oxygen to your tissues from your lungs.

What we've been talking about is *atomic diffusion* or *molecular diffusion*. Researchers figured out about 100 years ago that if you took out a sample for observation and suspended solid particles in a liquid, you would see them moving under the microscope. A researcher named Robert Brown saw this and decided to name the phenomenon after himself. He called it *Brownian motion*.

These two images show what it looks like:

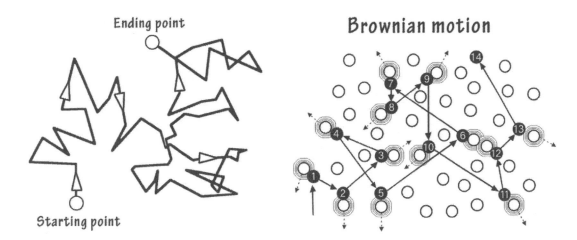

They also call Brownian motion *random walk motion* because it involves the random movement molecules that act like blind people bumping randomly into one another as they move around under the influence of energy or heat. Once the diffusion has completed so that the concentration of molecules is uniform, the movement doesn't change but the effect is that the molecules remain spreads out. The completion of the process leads to the *net diffusion* or net movement of molecules from an imbalance in concentration to an equilibrium state, where the molecular density doesn't change.

Remember this: *Nature likes states of low energy and high entropy. Don't you? Think of the state of your bedroom floor when you aren't expecting company. Lots of stuff strewn about is high entropy and there's no energy necessary to keep it that way. This is how nature intended things to be. Life defies nature by putting energy into building up things to a lower entropy state; this takes energy, but you can argue it's better than the reverse.*

Diffusion and Membranes

Diffusion happens in all biological systems, but it is made more interesting by the fact that membranes are involved. A *membrane* is a barrier that has a certain permeability. This can range from being totally permeable to not at all permeable. Permeability means the essentially the same thing as *permittivity*. A permeable membrane will permit more things to pass through than what it keeps out. Cell membranes are semi-permeable, which means some substances get a green light and others don't.

As we've mentioned, gases that are small like oxygen and carbon dioxide get green-lighted. So do most nonpolar molecules like lipids. Steroid hormones in the body (such as estrogen, testosterone, and cortisol) are lipid-based hormones, so they have no problem passing through the cell membrane. When you study *cell signaling*, you'll understand why there is no reason to have receptors for them on the surface of the cell. They do have receptors, but they are *inside* the cell instead.

The process of going through the cell membrane without any energy required, any carrier proteins needed, and from an area of high concentration to an area of low concentration is called *simple diffusion* or *passive diffusion* through the membrane. This process will always increase the entropy of the total system (inside and outside the cell). Because a membrane is involved, it isn't 100 percent the same as without any membrane; if a membrane is involved, the rate of diffusion depends on how easily the molecule slips through the membrane. The net diffusion will be the same, which means that given enough time, the membrane's effect will be negligible, and the diffusion of the molecules will mean equal concentrations of the molecule on both sides of the membrane.

This is what you should know about diffusion across a semipermeable membrane:

In this image, you can see that the concentration across the membrane is equal, but time is going to be a factor in how this happens. The timing will be different for each molecule and for each membrane type.

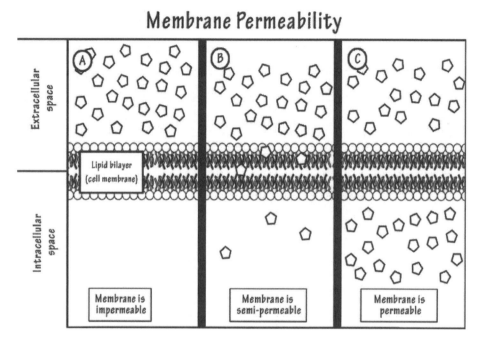

Size matters, as you can see by this image. If a molecule is big, even if it is nonpolar, it might not get through the membrane as all. Size, however, is less important than how hydrophobic or hydrophilic the molecule is that is trying to pass through the membrane.

Selectively Permeable Membrane

If you have a pure lipid bilayer without any proteins, this is what you'd see with regard to permeability:

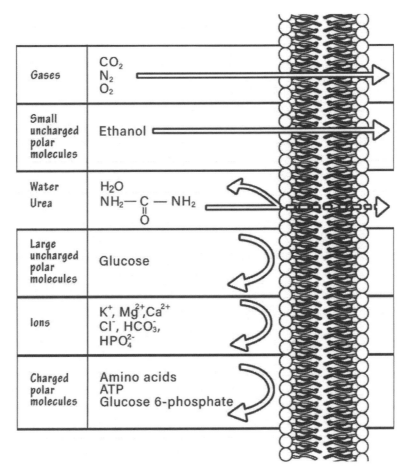

You can see how water and urea will get through a little bit but not very efficiently. There are special pores for water called *aquaporins* that get water through the cell membrane much faster than it can get through using simple diffusion alone. The aquaporin molecule allows water to get through the membrane via what's called *facilitated diffusion*. This just means it needs help to get across, even though no energy is expended in the process.

Osmosis is a Special Case

Osmosis involves a semipermeable membrane and molecules, but the main difference between diffusion and osmosis is that it involves solvents instead of solutes. In biological systems, this always means water. This is why we usually think of water when talking about osmosis, even though it could technically mean any other solvent in organic chemistry.

When we talk about osmotic *pressure,* we mean something kind of hard to explain. Technically, it is the amount of pressure you would need to apply to one side of the membrane so that no movement of water would happen across the membrane. It is easier to understand when you think of air pressure in the atmosphere as a force or pressure of its own. It is pushing down on the surface of any open container.

This next image is of a U-tube with a membrane at the base. As the water crosses through in order to equalize the concentration of water both sides, the water level on one side of the tube will rise. The difference in the pressure is called the *osmotic pressure*.

Osmosis

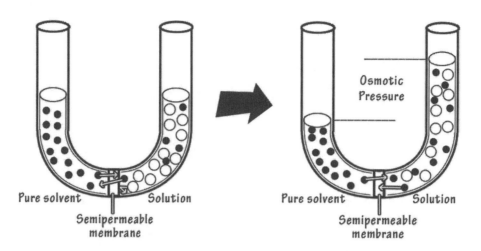

The force of air pressure pushing down on each side of the tube is the same, but there is more pressure on the right than on the left pushing back up on the air pressure. You can see that diffusion of solutes in the water did not happen. The number of molecules on either side is the same. The difference is that the right side was more concentrated than the left side. This means that the water passed through in order to equalize these concentrations, causing a rise in the water level.

Remember this:

• High concentrations of solute mean low concentrations of water.

• Low concentrations of solute mean high concentrations of water.

• Osmosis goes from where the concentration of water is high to where the concentration of water is low (from a dilute solution to a concentrated solution).

When it comes to cells, they like to be in solutions where the concentration of water is the same in the cell as outside the cell. These solutions are called *isotonic solutions. Hypertonic solutions* are too concentrated, so the cells will shrink in order to make the inside of the cell just as concentrated as the outside. The reverse is true in *hypotonic* or *dilute solutions*; the cell swells as water rushes in to equalize the total concentration of solutes.

When you talk about osmosis and osmotic pressure, you need to think differently than you do with diffusion in another way. With diffusion across a membrane, each solute is doing its own thing and will diffuse from its own high concentration to its own low concentration — regardless of what all the other solutes are doing. This isn't true for water. When it comes to osmosis and osmotic pressure, all dissolvable solutes are treated essentially the same.

Plants survive because of osmotic pressure inside the cell. The osmotic pressure in each plant cell is called its *turgor pressure*. The turgid plant cell is plump and juicy. In a leaf, it would look healthy and not wilting. If you don't water it, the leaf cells become flaccid and the leaf wilts. If you put the plant in a hypertonic solution of water, the water would rush out of the plant cell, and the cell would become *plasmolyzed*, or shrunken inside. The cell wall would give it some shape, but the cell inside would be a shrunken mess. This is what it looks like:

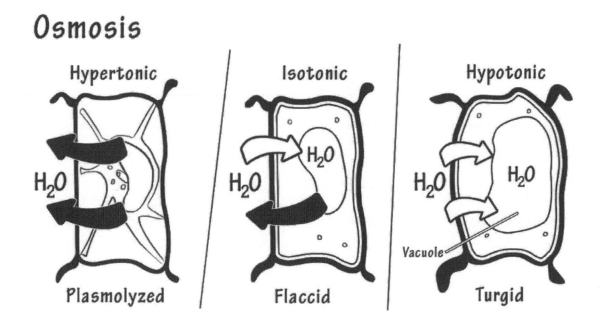

Osmosis and the pressures related to it are what cause roots to draw up water from the soil (although active transport of solutes happens to support this process). The solutes drawn up into the root cells of the plant make it concentrated with solutes. Water then gets sucked in through osmosis in order to compensate for the high solute concentration in the cell.

To Sum Things Up

Diffusion is the natural distancing that occurs when anything, including atoms and molecules, spread themselves out in order to increase the entropy in the system. In physics, you know that to reduce entropy, you have to add energy into the system. The same is true for biological systems. Diffusion across a membrane happens to cells that are permeable to the membrane but not to those things that can't cross it.

Osmosis is similar to diffusion, but it involves the solvent instead of solutes. Water tries to equalize its own concentration across a membrane that allows it to cross freely. It goes from a dilute solution to a concentrated solution, and it doesn't really matter what the solutes are as long as they can't cross the membrane themselves and are fully dissolved in the solution. Osmosis creates a certain osmotic pressure used in manufacturing and other technologies to do work.

CHAPTER 17:
PASSIVE AND ACTIVE TRANSPORT

The title of this chapter is a bit misleading because we've already talked about one form of *passive transport.* Technically, *simple diffusion* is a form of passive transport. The difference between diffusion and the types of passive transport we are going to talk about now is that the passive transport in this chapter involves some type of protein.

Passive Transport Using Proteins in the Membrane

As you can see so far, not many molecules get into or out of the cells by themselves; most require some type of protein channel to aid in these processes. Those that can pass through to some degree, like water and urea, still use protein channels to speed up the process. There are three main types of membrane proteins that are used in the transport of molecules across the membrane. Most of these are *transmembrane proteins* and are very picky about what types of molecules they let through. They aren't simple pores or holes that let anything of the right size pass through them. This image shows you the three protein types we will talk about:

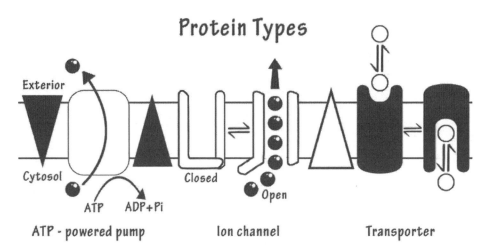

In the image, the first type is called the *ATP-powered pump.* This involves *active transport.* The meaning of "active" in this case is that it involves energy of some kind. The other two types are *passive transport*, which means no energy is involved. There are *channel proteins*, or *ion channels*, and *carrier proteins*, or *transporters*. Almost always, if a molecule must go against its concentration gradient or "uphill" from an area of low concentration to one of high concentration, energy into the system must occur.

Channel proteins usually are for the passage of ions or water from an area of high concentration to low concentration. These are like tubes or tunnels made from polar and hydrophilic proteins on the inside of the tube. Ions flow single-file but very quickly — up to 10^8 or 10,000,000 (ten million) molecules per second. An example of this type of channel is the potassium-specific channel protein that stays open in order to regulate the electrical charge differential across the cell membrane.

Transporters or carrier proteins move ions and other molecules across the membrane but at a slower rate than channel proteins. Up to a few molecules attach at a time, change the protein in a fundamental way, and slip through the membrane. This is less efficient than a channel protein, which is able to pass molecules through the membrane at a rate of about 100-10,000 molecules per minute.

Carrier proteins have three subtypes: *uniporters, symporters*, and *antiporters.* Uniporters involve one molecule type that crosses the membrane down its own concentration gradient without any need for added energy. Glucose and amino acids get into your cells this way.

Symporters also use two molecules at a time. Again, one of these travels across the membrane down its concentration gradient, using no energy to do this. This is coupled with another molecule traveling *in the same direction* to go against its concentration gradient. Once again, the energy gained from the first molecule's passage drives the second molecule's passage. The two reactions are coupled for maximum efficiency.

Antiporters use two molecules at a time. One of these crosses the membrane down its concentration gradient. This drives the ability for another molecule to cross the membrane *in the opposite direction* against its concentration gradient. The energy gained by the downhill part drives the uphill part.

Transporters

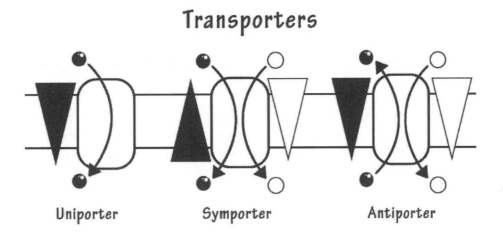

| Uniporter | Symporter | Antiporter |

More on Symporters

The simplest kind of transporter protein is the *uniporter.* There are known uniporters for *nucleosides, simple sugars*, and *amino acids* that are able to leave or enter simply based on the existing concentration gradient. Their purpose isn't to drive a process that is already energetically favorable but simply to make things happen faster. This is why it is called *facilitated transport.* This isn't the same thing as when an enzyme facilitates a reaction, but the effect is the same because these proteins essentially *catalyze* a process, not a chemical reaction.

Why isn't this the same thing as *simple diffusion* because there is a favorable concentration gradient in both of these?

1. Uniporters actually speed things up to become much faster than they would otherwise happen alone. Enzymes themselves aren't this fast.

2. Uniporters act specifically so that only a single molecule type or a related group of similar molecules are allowed to pass through them.
3. The passage of molecules depends on how many of these proteins there are. It does not depend on the total surface area of the cell membrane (which is true for simple diffusion).

The simple uniporter protein will be able to do the reverse of what it's designed to do, but only if the concentration gradient is reversed. This doesn't happen generally because the concentration gradient of the molecules using it hardly ever reverses itself. In the *intestinal epithelial cells* (lining the intestines), this process works well because there are a lot of these simple nutrient molecules in the intestinal lumen, and the molecules are whisked off for other purposes before the concentration differential could become reversed.

There is a uniporter molecule that helps all of our cells use glucose. It's called the *GLUT1 transporter*. It allows cells to take up glucose from the bloodstream and put it into the cells. This is an interesting protein that has an opening for glucose. As it takes glucose up, it is open toward the outside of the cell but, as glucose is spit out onto the interior of the cell, the opening faces the inside of the cell. This is what it looks like when it does this:

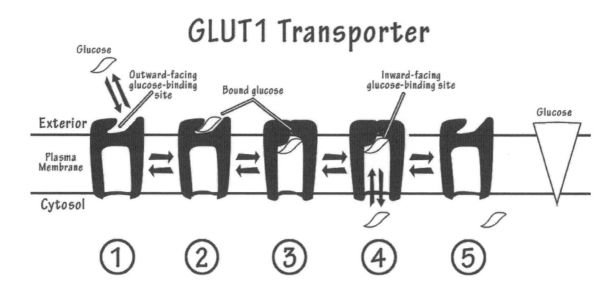

The way this uniporter works is interesting too. You would think that it would work faster if the glucose concentration in the cell was zero, but this isn't what happens. It works best if there is some glucose in the cell's interior that helps flip the open side of the protein to the outside. If there is no glucose in the cell from the beginning, this flipping doesn't happen as rapidly, so the transport is slower.

The reason this protein keeps working in the forward direction is that glucose quickly gets consumed for fuel, essentially leaving its own concentration gradient. It turns into *glucose-1-phosphate* instead so that the process transporting glucose across the membrane in an inward direction always moves forward. The affinity for glucose is so specific that the mirror-image isomer of the glucose we use (which is only D-glucose) will not pass through the GLUT1 uniporter much at all. Other D-isomers of simple sugars like *mannose* and *galactose* will be taken up to some degree, but the L-isomers won't pass through the membrane.

Fun Factoid: *Isomers called* enantiomers *are mirror images of each other. Their structure is chemically the same but their arrangement in organic chemistry terms is different. These are true mirror images that make the biggest difference in enzymatic reactions and protein receptors because these proteins have receptors that can tell the difference between these mirror images. If a molecule can have a unique mirror image, it is called a* chiral molecule. *This is what it looks like:*

Chiral molecule

More on Antiporters and Symporters

Symporters and *antiporters* are mainly for small molecules and ions so that they can travel against their concentration gradients. This type of movement is *facilitated transport*, but it can also be called *cotransport.* Most of the time, the driving force is that of sodium crossing the membrane in an energetically favorable way (down its concentration gradient).

The reason this works to help other molecules cross against their concentration gradient isn't so much because of the kind of energy you see put in by the breakdown of ATP. Even when that happens, it is isn't like the reaction is some kind of battery that electrifies the second reaction or process. For example, when glucose attaches to the carrier protein, it helps the second molecule also attach to same protein. When glucose does its thing naturally by going down its concentration gradient, the other process — that of sending the second molecule against its concentration gradient — is essentially obligated to happen, too. It's this obligatory relationship that causes the energy needed to drive the process against what you would otherwise get without this trick of nature.

There is nothing intrinsically different between a symporter and an antiporter except that they drive processes in different directions in relation to the sodium ion (or another similar ion). Both are cotransporters and both work by obligating a second molecule to go against its concentration gradient.

One of the main ways amino acids and glucose enter the kidney tubule cells is from the interior of the gut to the inside of the body. They both use sodium ions to drive the process. In the case of glucose, the symporter is called the *two sodium/one glucose symporter* because it takes two sodium ions to allow one glucose molecule to cross over with them.

The driving force in this process is two-fold. First, the concentration gradient of sodium drives the reaction in addition to the electrical charge difference. If you remember, the charge is more negative inside the cell, so the sodium ions (which are positively charged) would naturally want to equalize that charge difference. These two things together really helped drive this process in the forward direction. This is what it looks like:

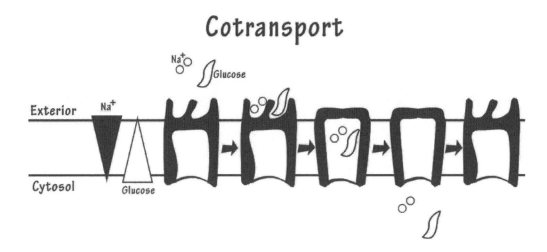

There is a sodium-linked antiporter, too. There is one called the *sodium-calcium antiporter*. It works in heart muscle cells to move calcium ions across membranes in the opposite direction from the sodium transport. It pumps the calcium out of the cell in order to keep its concentration as low as possible. In this case, it takes three sodium ions to force the transport of one calcium ion across the membrane.

Another antiporter is called the *AE1 protein* that regulates the ability of carbon dioxide to cross the membrane of red blood cells. In this case, it isn't the CO2 itself that is involved, but the bicarbonate ion. The process involves chloride as the driving force behind the reverse transport of bicarbonate across the red blood cell membrane. Both are negative ions.

Carbon dioxide as a gas really doesn't dissolve well in blood, so its transport in blood depends on a red blood cell enzyme called *carbonic anhydrase*. This transforms carbon dioxide gas into bicarbonate in the red blood cell. This is how the carbon dioxide gas gets transported from the tissues to the lung alveoli (air sacs). It turns out that it works in opposite directions, depending on where the reactions are happening. Take a look at this image:

The whole process allows bicarbonate to leave the red blood cell in the tissues, driven by the fact that carbon dioxide enters the red cell, gets turned into bicarbonate, and leaves via the antiporter system. Notice how the reverse happens in the lungs so that carbon dioxide can leave the red blood cells (and your body too). This is genius and allows you to get rid of carbon dioxide as a waste product of metabolism more easily.

There are other antiporters in animal cells that are important in human physiology. One of these is called the *sodium bicarbonate-chloride cotransporter*, which gets rid of too much acidity inside the cell. Sodium and bicarbonate go one way and chloride goes the other way. Bicarbonate enters the cell, mixes with hydrogen ions in the cytoplasm (which is the *acidity*), and forms carbon dioxide and water. The carbon dioxide leaves the cell through diffusion and the cell is less acidic.

There is also a similar protein to the AE1 protein in red blood cells found in all cells of the body. This protein does the same thing as the AE1 protein but doesn't depend on sodium ions like the one that gets rid of acidity in the cell. Still, another antiporter is the sodium-proton antiporter that adds to the acidity inside a cell. The goal of all of these antiporters is to regulate the pH. You can imagine that the chloride-bicarbonate antiporter lowers the pH of the cell if necessary, while the sodium-proton pump has exactly the opposite effect.

Active Transport with ATPase Pumps

True *active transport* means energy is necessary to drive the process; for the cell, this almost always means the energy must come from ATP. There are four known types of ATPase pumps in the cell. All of them transport something across the membrane against its concentration gradient using the energy derived from the hydrolysis of ATP. These four main types are called *P, F, V,* and *ABC classes.* The first three transport only ions, while the ABC ATPases are a group or superfamily of transporter molecules that can transport small nonionic molecules. This is what they look like:

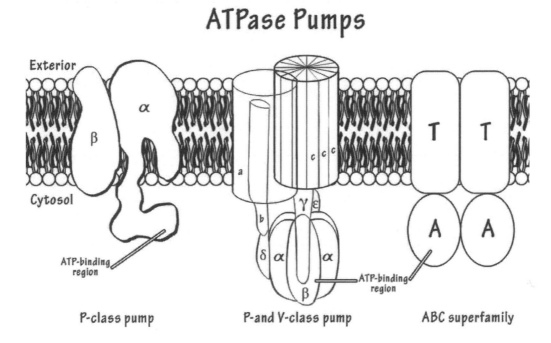

Here are the similarities and differences between the classes of ATPases:

- **P Class** — This class of proteins can transport hydrogen, sodium, potassium, and calcium ions. This is the class of the typical *sodium-potassium ATPase pump* used by all cells that is so important in establishing the inner cell environment. It is also the kind that pumps calcium into the sarcoplasmic reticulum of muscle cells in order to stop muscle contraction. Some of your stomach cells use this type to pump hydrogen ions into the stomach so it can be as acidic as it needs to be.
- **F Class** — This class only transports hydrogen ions across a membrane. This is the kind found in bacteria and in both *mitochondria* and *chloroplasts. F* and *V classes* do different things, but they are structurally similar to one another. It probably means they're related (evolutionarily speaking).
- **V Class** — This class only transports hydrogen ions across a membrane. It works in plants to make the vacuoles and lysosomes so acidic.

- **ABC Class** — This is a diverse group of protein pumps with 100 known different proteins that could possibly do the same thing (depending on the organism). These can transport small sugars, ions, small polysaccharides, and smaller polypeptides or proteins. The hydrolysis of ATP happens inside the cell and is always coupled with the transport of one or more molecules across the membrane. This coupling means that the energy from the process of ATP isn't wasted at all; it is a very efficient system.

The sodium-potassium ATPase is the most important ATPase pump in all cells, taking up as much as half the total cellular energy of the cell, depending on the cell type. This is the pump that maintains the difference in electrical potential across the cell membrane.

The process allows three sodium ions to leave the cell and two potassium ions to enter the cell by *hydrolyzing* (breaking up) just one molecule of ATP. After the process is over, the pump will be available for the process to start all over again.

Other ATPase pumps, such as the V and F pumps, have a unique mechanism as well. Their job is always to create an acidic side and a less acidic side across a membrane. When hydrogen ions get pumped into a lysosome, for example, other things also need to occur. If this happened in isolation, there would be a huge electrical charge difference inside the lysosome. The ATPase pump takes care of this by allowing chloride ions to pass into the lysosome as well (in order to keep electrical neutrality). In the stomach, H+ ions enter the stomach, while potassium ions flow the other way in order to keep everything electrically neutral here too.

Fun Factoid: One important ABC pump is called the chloride-channel protein. It is found in your lungs, pancreas, and sweat glands, among other areas. This might not seem important to you unless you have cystic fibrosis. In these people, a broken gene called the CFTR gene codes for this pump protein. Obviously, it doesn't work in a person with cystic fibrosis, so their secretions are very thick and they have lifelong problems with their lungs, pancreatic function, reproductive cells, and sweat glands.

To Sum Things Up

Most things cross the cell membrane through the action of proteins. There are ion channels and carrier proteins that allow molecules to cross through the membrane with or without energy needed. With facilitated transport, there is no ATP energy involved. Some, however, require an ion (usually sodium) to cross the membrane downhill with respect to its concentration gradient just so another molecule can go in an uphill fashion against its concentration gradient.

Active transport usually involves the hydrolysis of ATP in order to cause ions or molecules to cross through the cell membrane or other cellular membranes in unique ways. Some of these will cause an electrical gradient to happen across the membrane (along with an ion gradient), while others allow ions or molecules to pass through the membrane with mechanisms to maintain electrical neutrality.

CHAPTER 18:
BULK TRANSPORT OF
MOLECULES ACROSS A MEMBRANE

While it seems like some of these processes work very quickly to get molecules into and out of the cell, there are times when the cell needs to bring in a lot of molecules at once. This means the cell needs a way to have bulk transport of molecules. Instead of a steady drip of molecules crossing the membrane, a lot of molecules get dumped across the membrane at once. Sometimes, it involves molecules of the same type but often it doesn't.

There are processes that allow large numbers of molecules to enter the cell and other processes that do the exact opposite. Taking molecules into the cell this way is called *endocytosis*, while ejecting molecules out of the cell is called *exocytosis.* Let's talk about how each of these works.

Endocytosis

Endocytosis, or the bulk transport of molecules or other substances into the cell, doesn't happen randomly. The cell has ingenious ways of deciding when and how to do this. We've talked about *phagocytosis*, or "cell-eating." This is how bacteria and other pathogens get taken up into immune cells in the body so they can be destroyed effectively by these cells. This isn't a choosy thing. It takes receptors to detect what needs to be sucked up into the cell and what doesn't.

Pinocytosis, or "cell-drinking," is much less complicated than phagocytosis and not nearly as discriminating. Nearly every cell carries out this process. It involves the intake of a watery solution from outside the cell (packed inside a tiny vesicle); the vesicle membrane combines with the cell membrane and discharges the contents of the vesicle into the cell cytoplasm. This is what *phagocytosis, pinocytosis*, and *receptor-mediated* endocytosis look like:

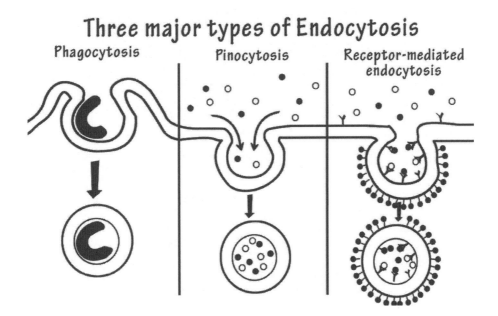

Three major types of Endocytosis

Phagocytosis   Pinocytosis   Receptor-mediated endocytosis

Receptor-mediated endocytosis is the kind that needs a signal. The signaling molecule is called the *ligand.* Once the ligand binds to the cell surface receptor, it triggers uptake of extracellular contents into a vesicle that then enters the cell. The complex of ligand and receptor often gets taken up into the vesicle at the same time. Some ligands that do this are cholesterol, iron, insulin, and other protein-based hormones.

How Does Endocytosis Work?

There are different ways that endocytosis can work. One major way this happens involves a protein on the inside of the cell membrane called *clathrin.* This protein must be present in order for the invagination to occur in endocytosis. The clathrin allows pits to be formed on the inside of the membrane; these pits get deeper and deeper until the whole thing comes together as a vesicle inside the cell. This image shows you how it happens, in general:

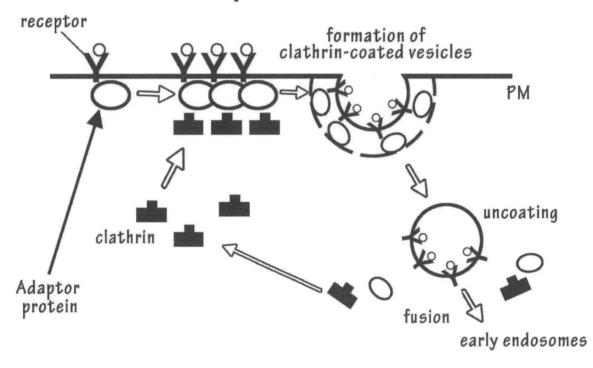

Once the vesicle gets inside the cell, it's called an *endosome.* These clathrin-coated pits are clustered under the cell membrane beneath receptors. The receptor binds to the ligand, starting the process (which requires clathrin). The clathrin proteins interact with one another in a regulated way so that the pit gets deeper and expands to form the endosome. That endosome is initially coated on the outside with the clathrin protein and its adaptor; eventually, these drop off and the endosome is relatively naked afterward.

One Example of Endocytosis Involves Cholesterol

Have you ever wondered what happens to cholesterol particles after you eat them? They aren't soluble in blood or plasma, so they need some soluble carrier molecule to help this happen. One of these carrier molecules is called *low-density lipoprotein*, or *LDL*. (No, your LDL is not the cholesterol itself; it's just the carrier molecule for your cholesterol.) LDL looks like this:

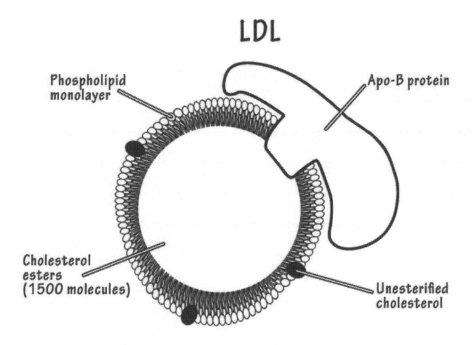

The protein is a sphere that has a large protein embedded in, called *Apo-B*, or *apolipoprotein B*. Your cells have receptors on them that bind to the LDL as a ligand, which allows you to take up the LDL complex, along with the cholesterol it carries. The LDL itself doesn't stay in the cell after it is endocytosed into the cell's interior. The molecules go to the lysosomes in order to be degraded; the cholesterol is separated from the protein and used by the cell. Most of it goes back into your cell membranes, while some gets stored into lipid droplets inside the cell.

Fun Factoid: Some people are born with a mutated LDL-receptor protein. It means that the receptors they have don't bind to the LDL molecule and cholesterol can't enter the cells. The end result is very high cholesterol levels and diseases like atherosclerosis, heart attacks, and strokes. The disease is called familial hypercholesterolemia and is a common inherited type of high cholesterol disease.

Exocytosis

Exocytosis, or "cell leaving," in large numbers is also regulated by the cell. Some cells do this a lot as part of their job. These include nerve cells that release neurotransmitters in order to signal other nerve or muscle cells.

There are other processes involving exocytosis in a cell that don't involve the bulk transport of a single molecule (as is true when neurotransmitter molecules are released by a cell). When molecules get packaged by the Golgi apparatus, for example, many of these leave the cell through exocytosis.

The Golgi apparatus sends out vesicles with molecules in them out of the trans face of the organelle. These are pre-packaged in tiny vesicles. This is what the process looks like:

Exocytosis

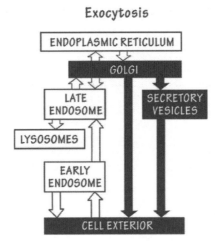

As you can see, vesicles leave the Golgi apparatus; they can be destined for secretion or be sent to the lysosomes for destruction. Hormones, neurotransmitters, and many digestive proteins get packaged in large numbers and are stored in the cell cytoplasm for mass release when needed. This leads to three pathways: 1) the *constitutive secretory pathway*, which isn't regulated, 2) the *lysosomal pathway*, and 3) the *regulated secretory pathway* for stored molecules like neurotransmitters.

When the Golgi apparatus does its job, there are three main destinations for the products they release: 1) the lysosomes, 2) the secretory vesicles for storage, and 3) the cell membrane for immediate release. Those tagged to go to the lysosomes have a sugar coating of *mannose-6-phosphate*. This sugar coating is the "message" that tells the vesicle to head to the lysosome. If there is no signaling message on the vesicle, it automatically goes to the cell membrane for release. Let's look at the steps involved for these pathways:

1) **The secretory vesicles bud out of the Golgi apparatus.** — When we say, "secretory vesicles," we mean those that are stored for later use and molecules like digestive enzymes or neurotransmitters. There are signals from outside the cell that tell these vesicles to leave the cell through exocytosis. These vesicles usually contain just one molecule type per vesicle. Here's how these vesicles are made:

Vesicles

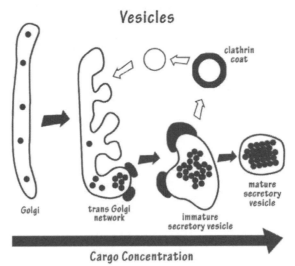

These secretory vesicles travel to cluster near the cell membrane so that when the signal is given, the vesicles can discharge many of the same molecules at once. The signal in nerve cells, for example, is an electrical charge at the tip of the nerve cell. It causes calcium to flood the area; this is the trigger for exocytosis to happen. The collection of secretory vesicles near the cell membrane is called *docking.* Besides calcium, proteins called *SNAREs* help the vesicles to fuse at the appointed time.

2) **The vesicle membranes get removed from the cell membrane.** — Once exocytosis happens, the cell membrane will recycle the lipids involved (that came from the Golgi apparatus). Endocytosis happens rapidly after the contents of the Golgi vesicles are dumped out of the cell so that the cell membrane doesn't just keep growing over time as lipids are added to it each time exocytosis happens.

3) **Regulated exocytosis also involves budding out of the Golgi apparatus.** — An example of this is the release of *histamine* (the molecule that makes you itch) by *mast cells*. If you are allergic to something, for example, a signal gets to the mast cell and histamine is released in large numbers. If you get a lot of histamine released at once, you might get hives.

4) **There is a way to have these vesicles leave in specific places.** — Imagine a cell lining the intestines filled with digestive enzymes. If these are to be released into the gut, they must leave out of the face of the cell that actually faces the gut lumen. The Golgi apparatus adds proteins on the vesicle surface so it will only bind and discharge itself on the face of the cell that will respond to these protein-based cell signals. Here's how this happens:

Vesicles Transport

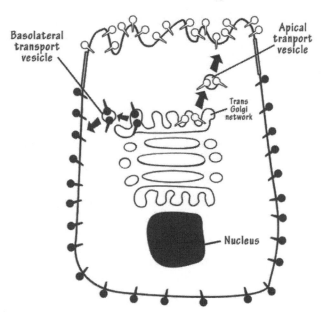

5) **Nerve cells have two different ways of doing exocytosis.** — Nerve cells and a few endocrine cells have two types of these secretory vesicles. There are typical vesicles that are densely packed with proteins or other molecules. There are also tiny secretory vesicles for small neurotransmitter molecules. These tiny vesicles are discharged rapidly once the signal is given for this to happen. Some neurons fire off a signal 1,000 times per second, so the process must be fast and efficient. It takes tons of tiny vesicles already docked and ready to be discharged

at a moment's notice. Because everything must be recycled so quickly, many get recycled very close to the cell membrane, so there isn't much time before new, replenished vesicles are ready to go. Here's how this happens:

As you can see, both endocytosis and exocytosis are necessary to have this important process happen over and over again in the nerve cell.

Exocytosis

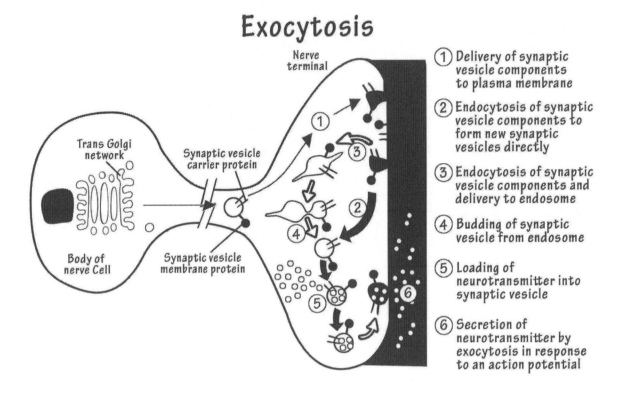

1. Delivery of synaptic vesicle components to plasma membrane
2. Endocytosis of synaptic vesicle components to form new synaptic vesicles directly
3. Endocytosis of synaptic vesicle components and delivery to endosome
4. Budding of synaptic vesicle from endosome
5. Loading of neurotransmitter into synaptic vesicle
6. Secretion of neurotransmitter by exocytosis in response to an action potential

To Sum Things Up

Endocytosis and exocytosis are both bulk transport mechanisms in the cell. With endocytosis, much of it happens after a receptor attaches to a ligand (receives a signal), triggering the clathrin on the inside of the cell membrane to form a deepening pit where the vesicle is ultimately formed.

Exocytosis involves leaving the cell using a vesicle. Almost all of these come from the Golgi apparatus, which prepackages these molecules into tagged vesicles that will "know" exactly where to go. Some that aren't tagged just go to the membrane to be exocytosed from the cell. Others go to the lysosomes for destruction of the molecules. Still others are called "secretory vesicles" that wait for a signal before fusing with the cell membrane.

CHAPTER 19:
CELL SIGNALING

All types of cells do some kind of cell signaling — even *prokaryotic*, single-celled bacteria. *Signaling* is basically the main way cells communicate with one another so they can work together. This can get very complex, considering each cell can have hundreds of different signals it must respond to in order to change its behavior according to the environment.

Some signaling happens over a short distance. This is called *paracrine signaling,* meaning one cell communicates with another cell in the same vicinity. There is also signaling where a cell communicates with itself, called *autocrine signaling*, as well as *juxtracrine signaling* to the cell next to the signaling cell through gap junctions between them. In people and complex organisms, you can also see *endocrine signaling*, which is when a signal goes a long way through the circulatory system before reaching its target cell using cell signaling molecules better known as *hormones.* This is what it looks like:

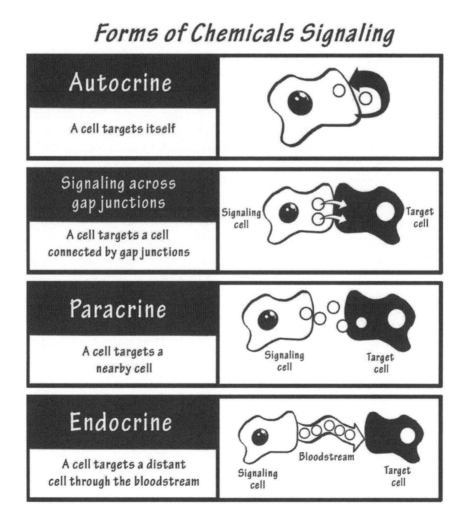

The target cell is also called the *effector cell* because when it is targeted, it has an effect on the cell. It takes action such as changing its metabolism, beginning cell division, enhancing cell growth, or changing the direction of its motility.

You might think that the communication starts and ends when a signaling cell gets its message to the target cell, but this is far from the truth. Once the target cell is reached and the ligand or hormone attaches to the receptor, the process keeps going so that the external signal becomes an internal message that causes the cell to do something. A message that only reaches the cell surface wouldn't do much to the cell unless the message got inside somehow. Fortunately, cells have ingenious ways of doing this.

You can think of it as two kinds of messengers. There is a *main messenger* or *extracellular messenger* that usually acts entirely outside the cell. There are many of these messenger molecules that go by different names. You can see these kinds of messenger molecules called *growth factors, hormones, cytokines*, and *neurotransmitters.*

There are also intracellular messengers called *second messengers* that act only within the cell itself. We will talk about some of these with names like *cyclic AMP, cyclic GMP, calcium,* and *phosphatidyl inositol*. Nitric oxide is cool because it is both a gaseous extracellular signaling molecule and an intracellular messenger molecule.

How Does Cell Signaling Work?

The process of cell signaling depends greatly on the message or signaling molecule. If the signaling molecule is polar and can't cross the cell membrane, of course it's going to need a receptor on the target cell surface in order for it to have any effect on the cell at all.

In addition, if you have a perfectly good signaling molecule but the target cell has no receptor for it, the signaling molecule will be ignored. If the target cell has a few receptors for the signaling molecule, there might be a weak effect on the cell, and if there are lots of receptors on the target cell, the effect could be much greater.

If a signaling molecule is nonpolar like a steroid hormone (made from lipids) or if it is a gas like nitric oxide, these can easily get into the cell. The receptors for these hormones and signaling molecules are located inside the cell. The effect might be no different than if it were a polar molecule with a surface receptor, but the part that takes the external message and turns it into an internal message is skipped altogether.

These are the basics of cell signaling in a nutshell:

Cell Signaling

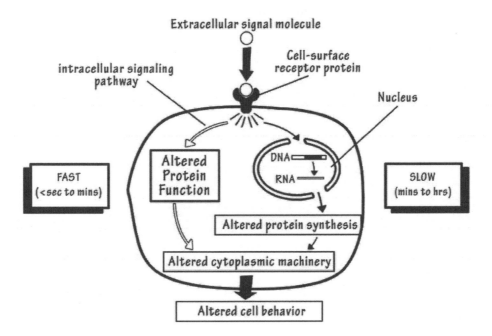

You see how the cell gets acted upon internally just by having something attach to it externally. There are mechanisms in place in each cell (and associated with the different receptors) that allow the cell to change in fundamental ways once the external message gets translated into an internal one. The effect can be fast or slow, depending on the outcome. For example, if the effect is to turn a gene on inside the cell, there could be a big change even though it isn't immediate.

Surface Receptors are the Beginning

The three main cell surface receptors are:

- G protein-coupled receptors
- Enzyme-linked receptors
- Ligand-gated ion channels

Researchers have figured out which kinds of hormones and other signaling molecules are involved with what receptors. For example, there are a lot of hormones and other molecules that attach to the *G protein-coupled receptor*, or *GPCR*. These include some you might heard of like *histamine, dopamine,* most pituitary gland hormones, and *adrenaline.*

These are the three types of membrane surface receptors:

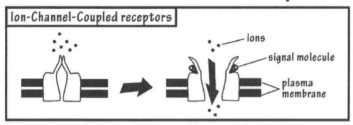

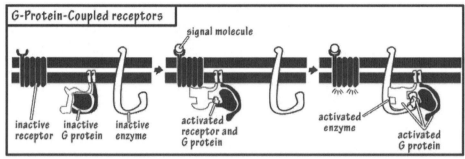

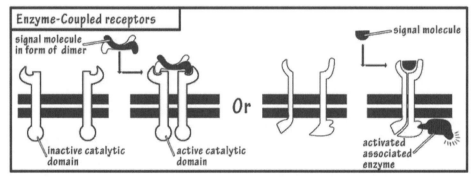

Let's look at the G protein-coupled receptors first. These are giant proteins that cross the entirety of the membrane from the outside to the cell's interior. Up close they look like this:

G Protein - Coupled Receptors

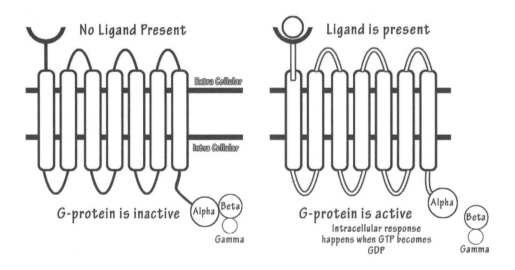

Once the receptor is bound to the ligand, the G protein inside the membrane gets activated. This takes energy from GTP. The GTP is a lot like ATP but with guanine instead of adenine. The reaction is the same so that GTP becomes GDP, releasing phosphate and energy that cleaves off a couple of subunits from the main protein. This will lead to a response inside the cell.

Another receptor type is called the *enzyme-linked receptor.* Here is what those look like:

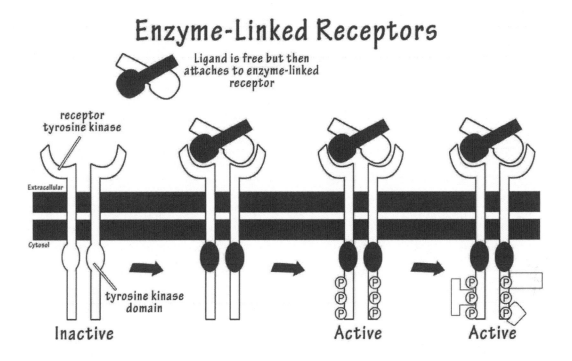

The example shown is called *receptor tyrosine kinase*. It works like this: You have a protein that spans the membrane. The outside part is the receptor that binds to the ligand molecule. On the inside is a pair domains or segments called *tyrosine kinase domains.* These get activated when the receptor gets bound but only if both ATP energy and the enzyme called *tyrosine kinase* are available to add phosphate groups to the part of the protein that has tyrosine amino acids on it. Once the phosphate groups are attached, molecules can stick to them much easier. This goes on to affect the cell in ways we will talk about soon.

Ligand-gated ion channels are easy to understand. Here's what these look like:

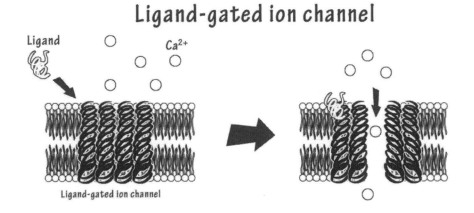

The idea is simple. The ion channel is a protein that spans the cell membrane. It is closed (usually) as long as no ligand is present. If a ligand attaches to the channel protein receptor site, the channel opens up and the ions (like calcium, sodium, and potassium) can flood the inside of the cell. This will affect the cell in different ways.

Second Messengers Act on the Inside

Second receptors are like the middle managers in the cell. They take the message they get from the cell exterior that has been passed to the inside of the cell and do something about it. These are small yet powerful molecules. Some examples of these are *cyclic AMP, cyclic GMP, inositol triphosphate (or IP3), calcium ions,* and *nitric oxide.*

There are only a few known second messengers that can have many different effects depending on the cell. Let's look at a couple of these to see what they do. Cyclic AMP, or *cAMP*, is a small molecule made out of ATP. When a receptor gets activated, it turns on an inside enzyme called *adenylyl cyclase* that turns ATP into cyclic AMP. It looks like this:

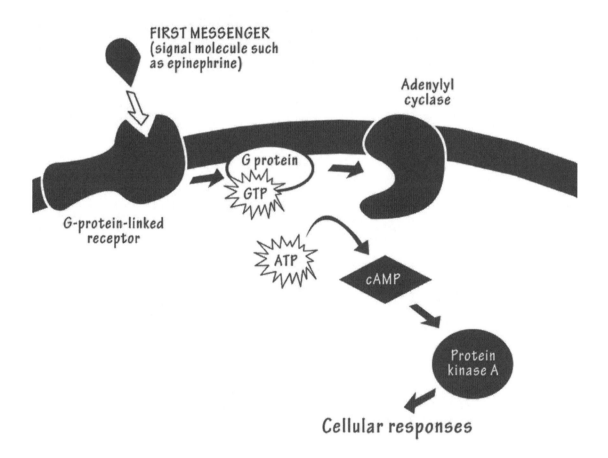

See how the ligand (epinephrine) binds to the G protein-coupled receptor, causing GTP to fuel the reaction that starts with ATP and goes to cAMP (by activating adenylyl cyclase)? The cAMP has multiple actions on the cell, most of which involve different biochemical pathways that affect the cell's metabolism. One of these is to activate *protein kinase A* in the cell that goes on to doing other things in the cell.

Another common second messenger is called *IP3*, or *inositol triphosphate.* It works like this:

Inositol Trisphosphate (IP$_3$) Pathway

It is a bit more complicated but just follow this: 1) a membrane receptor gets activated, which activates a nearby molecule in the membrane called *phospholipase C.* This is an enzyme. 2) Phospholipase C, or PLC, acts on another nearby membrane molecule called *PIP2*, or *phosphatidylinositol.* 3) A molecule of IP3, or inositol triphosphate, breaks off and triggers the release of calcium ions hiding inside vesicles called *endosomes* (inside the endoplasmic reticulum). 4) Calcium floods the cytoplasm and itself acts on yet another molecule called *PKC*, or *protein kinase C.* This enzyme works best when *DAG*, or *diacyl glycerol*, is attached as well.

There are 15 total protein kinase C types in humans. They each do different things. Remember that all kinases add phosphate groups to something. This is what PKC does in the cell. Many different molecules and even other enzymes function best when they are phosphorylated. Others though, might get turned off if they are phosphorylated. This is the job of protein kinase C. The end result is that a lot of different cell activities get turned on or off based on whether or not one of the many protein kinase C enzymes has activated the right enzymes needed to do the job.

It's amazing. You start with a simple receptor that gets bound to a ligand outside the cell. The message gets passed to the inside of the cell, where secondary messengers take over. These cause many different biochemical pathways to happen inside the cell.

What is the end result? A cell can grow, divide, metabolize differently, make new proteins, migrate somewhere, or even die. This is the impact of cell signaling. You can see that the impact is big and that this is basically how all cellular processes are regulated.

Fun Factoid: How does caffeine work? It turns out that it binds to the adenosine receptors in the brain, blocking them from being able to be bound by their normal ligand, adenosine. Adenosine normally causes your brain cells to feel sleepy. With caffeine on board, adenosine can't do this, so you feel less sleepy and more stimulated. In the pituitary gland, it causes hormones to trigger your adrenal glands to make more adrenaline so that you feel more attentive and energetic — until it wears off and you need another jolt of caffeine.

Regulation Through Feedback Systems

There are feedback systems that tell the signaling cell to stop sending more signaling molecules. Most of these are "negative" feedback systems. The idea is that, if a target cell is triggered to make a certain product as a result of being stimulated by a ligand, it should be able to turn it off somehow. It does this by releasing the product it made (or by sending a different signal that says, "enough already!") This product will act as its own signaling molecule that turns off the original signaling cell.

Another option is that the target cell will decide it doesn't need any more signaling, so it will bind to its own receptor sites, deactivating it temporarily. Without any available receptors, the target cell will simply ignore any more of the signaling molecules. This goes on until the cell needs to be more receptive to another signal. The molecules that deactivated the receptor drop off so the cell can be more responsive again. Negative feedback can happen quickly or slowly; it can also last a short period of time or a very long period of time.

There are also positive feedback systems where an action gets amplified over time. A little bit of a response happens inside the cell, which causes the receptors to be even more responsive to the ligand. There is a thing called *upregulating* too, where the receptor activity gets enhanced or upregulated simply by making more of them. This is what *upregulation* and *downregulation* of receptors look like:

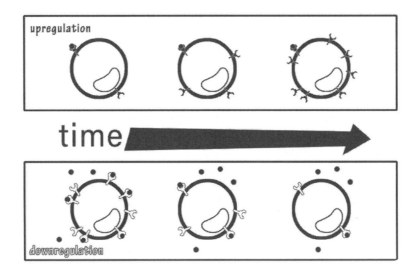

What About Intracellular Receptors?

Lipid-based and nonpolar ligands can skip the step involving an extracellular receptor altogether. They are hydrophobic, so they slip through the cell membrane and get into the cell nucleus. There are receptors there that, when activated with their ligand, will bind directly to DNA, activating a gene to start the process of making whatever protein it's been signaled to make. This is a slower process in general, but it gets the job done nicely.

Most of the ligands that do this are called *steroid hormones*, although there are others. Examples you see in the human body are *thyroid hormones, vitamin D, retinoids, cortisol*, and almost all of the reproductive hormones like *testosterone, estrogen,* and *progesterone.*

Basically, it works like this:

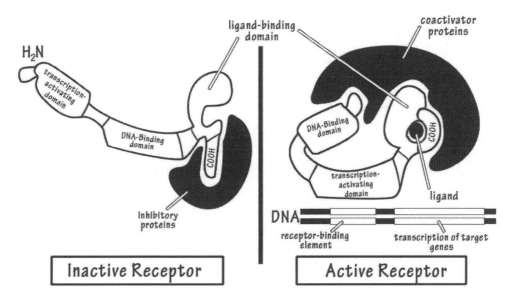

The molecule on the left is the nuclear receptor. On the right, it gets bound by the ligand (usually a steroid hormone or other hydrophobic molecule) plus one or more coactivator proteins. These all attach to the DNA strand, turning on the gene it has attached to. From here, the gene starts transcribing messenger RNA, which goes on to make a new batch of proteins that perform many functions inside the cell.

To Sum Things Up

Cell signaling involves ways cells communicate with each other. Cells can communicate through *autocrine signaling, juxtracrine signaling, paracrine signaling,* or *endocrine signaling.* All of these involve some type of signaling cell that makes a signaling molecule, or *ligand.* Ligands can be polar or nonpolar.

The polar signaling cells are also called *first messengers.* They bind to cell surface receptors, triggering them to pass a message into the cell. From here, the *second messengers* take over. These will cause many different biochemical pathways to occur in the cell that do dozens of different things that change the cell's behavior. Feedback systems (both positive and negative) will turn off the process or enhance it further, depending on the system.

Intracellular molecules also act as ligands if they can get through the membrane by themselves. They travel to the cell nucleus, where they bind to one or more protein receptors. Once this happens, a gene gets activated (or turned off), affecting the protein production in the cell. The cell's behaviors change based on what new proteins are made or on what proteins don't get made as a result of the ligand's actions.

SECTION SIX:
CELL RESPIRATION AND PHOTOSYNTHESIS

This next session involves a lot of biochemistry. We will talk about the most important biochemical processes any cell will ever participate in. These are the processes that give the cell the necessary energy to drive nearly every other biochemical pathway. *Cellular respiration* is what takes the fuel you eat and turns it into ATP energy and waste products. Without this cellular engine and the fuel you take in, nothing in the cell can take place.

We will also talk about *photosynthesis*, which takes light energy and turns it into usable glucose that animals and other organisms utilize in a cycle that starts with carbon dioxide in the atmosphere and ends with carbon dioxide (and oxygen) in the atmosphere. The process goes from photosynthetic organisms, or *phototrophs*, to *heterotrophs* that use the glucose made by plants in order to generate fuel and give the carbon dioxide back the plants. These processes participate in what's called the *carbon cycle*, which looks like this:

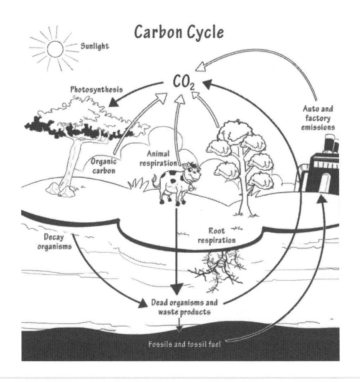

CHAPTER 20:
OXIDATION AND REDUCTION

Let's go back to oxidation and reduction again. Metabolism on a molecular scale is really a balance between oxidative and reductive processes.

As energy in fuel is consumed, ATP energy is made. Remember that these are called *catabolic processes*, or *catabolism*. The entire process of turning glucose into CO2 and water is called *oxidizing* glucose. It takes oxygen to do this fully; the oxygen gets reduced at the same time in a reaction that looks like this:

$$C_6H_{12}O_6 + 6\ O_2 \rightarrow 6\ CO_2 + 6\ H_2O + energy$$

Photosynthesis isn't biochemically the opposite of aerobic respiration in plants and other photosynthetic organisms, but the effect is the same. It starts with CO2 and "fixes it" to make glucose, which is a reduction reaction. The pigment, called *RH* in this next equation, is reduced as part of the process. This is what it looks like:

$$6\ CO_2 + 12\ RH_{red} + energy \rightarrow C_6H_{12}O_6 + 12\ R_{ox} + 3\ Oxygen$$
$$\text{(or equivalent)}$$

You should know that not all organisms on Earth go through this. In the deep ocean, for example, there are bacteria that use hydrogen sulfide in order to make glucose from CO2.

It's All About the Electrons

Remember that oxidation and reduction reactions must go together. How do you know this? It's because both of these types of reactions involve either donating or receiving electrons. If you have a reaction where electrons are donated, you need a receiving substance. Something always gets oxidized and gives up an electron and another thing accepts the electron and becomes reduced. Remind yourself of what these *redox* reactions look like with this image:

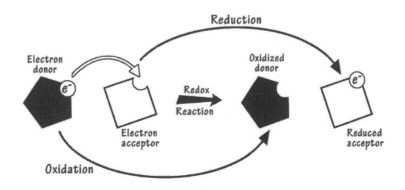

Remember this:

- An oxidizer wants electrons, so it takes them from another molecule in order to oxidize the molecule. In doing so, the oxidizer itself gets reduced.
- A reducer donates electrons (and hydrogen) in order to reduce another molecule. In doing so, this molecule itself gets oxidized.

In the cell, it looks like this: First, the cell gets hungry and eats a sugar molecule (usually glucose). The idea behind eating the glucose is to use it for energy. In order to do this, the cell has reactions in a pathway or pathways that chew up the glucose. As it does this, six carbons get broken down into two three-carbon molecules. These get further broken down into two-carbon molecules.

In most cells in humans, the process continues until there are six one-carbon molecules left for a total of six carbon dioxide molecules plus water to get rid of the hydrogen atoms. If this idea of "donating electrons" confuses you, you might want note that the carbon atoms on the glucose molecule have a mixture of hydrogen and oxygen molecules around them. (The chemical formula is C6H12O6.) The carbon atoms in CO2 have oxygen atoms around them and no carbon atoms. These are clearly more *oxygenated*, so this would be the oxidation part of the reaction.

In cases where CO2 is made out of glucose molecules, the reaction needs oxygen. The oxygen molecules start out with no hydrogen atoms or carbon atoms around them but later have hydrogen atoms around them in the form of H2O. This is less oxygenated, so it is the reducing part of the equation.

Putting this together, you get glucose being oxidized and oxygen being reduced. It looks a lot like this:

$$C_6H_{12}O_6 + 6O_2 \rightarrow 6CO_2 + 6H_2O + energy$$

(becomes oxidized; becomes reduced)

So, while you can think of how much each molecule is oxygenated in this type of redox reaction as being a measure of whether the molecule is oxidized or reduced, it really is about giving or accepting electrons. Any time an electron is donated, a hydrogen atom leaves with it and vice versa (in this reaction specifically).

The process of breaking down, or *oxidizing*, glucose is a catabolic reaction that will naturally give off energy. This energy is not random "energy" or anything like the energy you see when you plug an appliance into an electrical source. Instead, the energy goes straight into making ATP, which is every cell's major energy currency.

How much energy do you get? This has been calculated already. If glucose gets "combusted" into CO2 and water, it gives off 686 kilocalories per mole of glucose. This means that the change in *Gibbs free energy* is -686 kcal/mole. The whole thing looks like this:

C6H12O6+6O2→6CO2+6H2O

$\Delta G = -686kcal/mol$

Or, here is the same idea on an energy diagram:

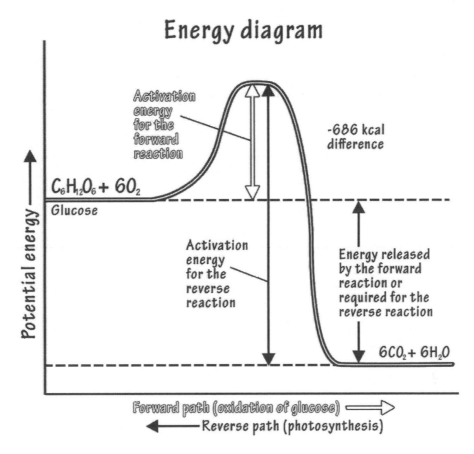

Instead of just combusting or "burning up" inside your cells, the energy is harvested by ADP to make ATP. Yes, it does give off heat, so you are definitely warmer than if these reactions hadn't happened, but that's because the reaction processes aren't perfect. Heat is given off because of this imperfection. If it were a perfect system, all of the energy would go to making ATP.

The whole thing isn't a waste of heat, though, because in warm-blooded people, you really need that heat to regulate your body temperature. The enzymes in your cells need the heat to function. Remember that enzymes are temperature-dependent; they need a certain heat range in order to function. Basically, it means the heat "wasted" is still important.

Redox in the Cell

The cell will participate in *redox reactions* all the time. The main redox situation, however, is the breakdown or catabolize of glucose and other molecules. In the next chapter, we'll talk about the three separate steps in this catabolism of nutrients that happen in most animal cells. If you'll recall, these include:

- Glycolysis
- Krebs cycle (also called *tricarboxylic acid cycle* or *citric acid cycle*)
- Oxidative phosphorylation (using the electron transport chain)

You put these things together and you accomplish the complete oxidation of glucose, the complete reduction of oxygen, and the making of ATP energy (plus heat).

You will also learn that ATP is made in two ways. The making of ATP from ADP is called *phosphorylation.* The two types of phosphorylation are 1) *substrate-level phosphorylation* and 2) *oxidative phosphorylation.* Substrate-level phosphorylation does not require oxygen, and the energy is made as part of a biochemical reaction. Oxidative phosphorylation uses oxygen and involves the electron transport chain. Of these, the electron transport chain makes much more ATP than does substrate-level phosphorylation.

In a nutshell, oxidative phosphorylation and substrate-level phosphorylation look like this:

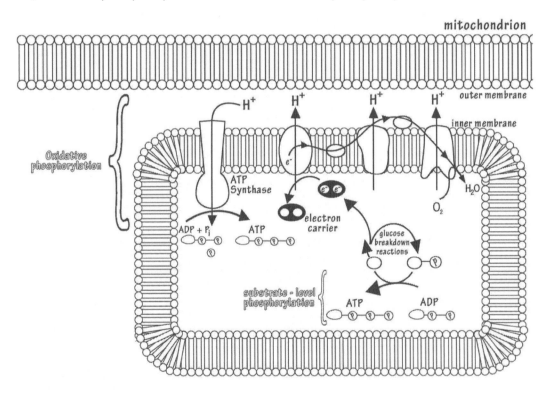

If this looks confusing, don't worry. It will make complete sense in the next few chapters.

Let's look at how substrate-level phosphorylation works in the cell. The concept of *substrate* means that a reaction is happening where a substrate goes to make a product. In the reaction, ADP is one of these substrates and ATP is an end product. A total of two ATP molecules per molecule of glucose get made this way (actually four get made but two are used up, so the net gain is two ATP molecules). The reaction looks a lot like this:

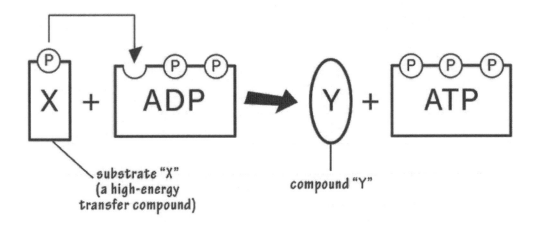

substrate "X"
(a high-energy
transfer compound)

compound "Y"

As you can see, a number of ATP molecules are made this way, but, as the total number made is really about 32-34 ATP molecules, the number made through substrate-level phosphorylation is comparatively small.

Oxidative phosphorylation is really where the action is when it comes to making ATP energy. This is much more complex but is really ingenious. It requires mitochondria to perform this process and mainly involves the inner mitochondrial membrane. It's in this membrane that enzymes participating in this process are embedded and where the *proton pump* provides the necessary energy for it to happen. This process makes a whopping 34 molecules per glucose burned.

As you will see in the next chapter, the maximum number of ATP molecules that could get made don't actually end up in the final tally. This is because the process isn't as efficient as it could be. Because this is no different from any engine or factory, the efficiency you see will not be 100 percent. In fact, some people have calculated the efficiency of these processes as being only about 38.3 percent.

The process of oxidative phosphorylation depends on something called *chemiosmosis* because it involves the passage of a chemical (hydrogen ions) through a membrane—which is loosely like real osmosis but doesn't involve water or a solvent (so it isn't really osmosis at all). Chemiosmosis looks like this:

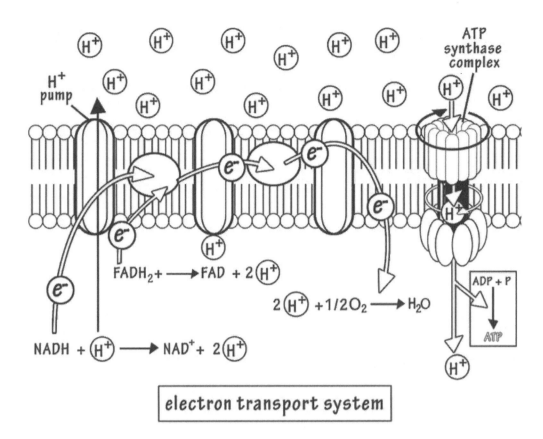

$$FADH_2 + \longrightarrow FAD + 2\,H^+$$

$$2\,H^+ + 1/2O_2 \longrightarrow H_2O$$

$$NADH + H^+ \longrightarrow NAD^+ + 2\,H^+$$

ADP + P

ATP

electron transport system

Notice the *chemical* or hydrogen ions on one side of the membrane and the channels through which they pass in a sort of chemical gradient. Because charged hydrogen ions are the chemicals involved, this is also an electrical gradient across the membrane. This electrical gradient drives the enzyme we've talked about — ATP synthase — to make ATP energy.

Electron Carriers in Cellular Redox Reactions

Electrons and hydrogen atoms don't just go from glucose to carbon to oxygen to make CO2 and water directly. This is a long process that involves many electron carriers. Some biologists call these *electron shuttles* because they catch electrons and hydrogen atoms but don't keep them long. They pass them down to another molecule like a bucket brigade. It would be the same as dumping water on a fire gallon by gallon rather than using a single large tank.

In the cell, the process of using electron carriers takes many smaller energy-reducing steps so that little by little, the reaction processes go downhill from an energy perspective. It looks like this:

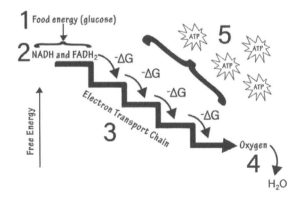

See how it is a stepwise reduction in energy but that the total energy is the sum of the different steps? Notice too how the molecules NAD and FADH2 are involved in the process. Let's look at these electron carrier molecules close up.

Both NAD+ (*nicotinamide adenine dinucleotide*) and FAD (*flavin adenine dinucleotide*) are known electron carriers that hold onto hydrogen atoms and electrons for a period of time before they go on to the *final electron receptor* molecule, which is oxygen. This is what NAD+ looks like and what happens to it when it becomes "reduced" temporarily to make NADH. Because these are reduced, they are turned into oxidizers themselves (because they have electrons now to give away).

FADH does the same thing but becomes FADH2 after it is reduced. Both of these molecules are made before the electron transport chain but then get used up in order to make ATP energy.

Fun Factoid: Mitochondrial diseases in humans are devastating. There are a lot of genetic disorders where mitochondria don't work. People with these problems have severe muscle weakness, poor coordination, and different kinds of brain dysfunction just because their energy-making factories are not working. Some adult diseases, such as type 2 diabetes, involve problems with the oxidizing efficiency of their mitochondria. Researchers are studying the best ways to treat these diseases.

To Sum Things Up

Cell metabolism — in particular, *cell catabolism* of nutrients — involves *redox reactions*. Redox stands for *reduction* and *oxidation*. It means that there are a pair of reactions where one gives up electrons as reducer to another molecule in order to reduce the molecule and another takes electrons from another molecule in order to oxidize it. These things must go together.

In cell metabolism and in catabolism, if a molecule gains a hydrogen ion, you can assume it is reduced. The reactions we will talk about next are the primary ones in the cell that oxidize glucose to make CO2 and reduce oxygen to make water. (Water is oxygen plus two hydrogen atoms, so you can see that it was actually reduced in the process.)

There are *electron shuttles*, or *electron carriers*, in cell systems that are made early on in the catabolism of glucose that will be used later in the process. The way they get used is to donate their electrons and hydrogen atoms in a series of reactions that ends up with oxygen taking these straggling hydrogen atoms and electrons. There, the process stops as no further oxidation can happen to carbon dioxide and no further reduction can happen to water. The end result, though, is more than 30 ATP molecules made for every glucose molecule burned up in this cellular energy factory process.

CHAPTER 21:
STEPS OF CELLULAR RESPIRATION

We've talked so much about this process of catabolizing glucose that you might think you know this completely, but there is a lot more to it than that. In most cases, you do not need to memorize each of the steps it takes for a glucose molecule to become carbon dioxide and water. It's probably better to understand that there are a lot of steps involved and that many of these steps contribute to the process of creating ATP energy.

Glycolysis is the Beginning

It all starts with a glucose molecule. This sugar is important to all cells in nature, and all cells undergo *glycolysis* (although a few have slight variations in this biochemical pathway). What this tells us is that evolutionarily speaking, this is a very old pathway that is necessary for life.

The take-home message for this process is this: One 6-carbon glucose molecule gets broken down into two 3-carbon molecules of *pyruvate*. In organisms that do not use oxygen, this is sort of where it ends, although they have creative ways of turning the pyruvate molecules into what we call *fermentation*. Organisms that quit at the fermentation stage make a few different end products you might be familiar with:

- Ethanol, which comes from organisms like yeasts that make the alcohol you drink
- Lactic acid, which is the byproduct made when your muscles are overused in the absence of oxygen (called *anaerobic metabolism*)
- Glycerol
- Acetone
- Monosodium glutamate
- Butyl alcohol
- Acetic acid
- Citric acid, some antibiotics, gluconic acid, riboflavin, and vitamin B12 — all made by certain molds that undergo fermentation

In all cells, the process of glycolysis takes place in the cytoplasm. Biochemically speaking, there are three separate stages, but there are two halves of the process if you think of it as the energy processes in the pathway. These two halves are the *investment half* and the *payoff half*. In the investment phase, the process requires the input of ATP energy. It's kind of like any investment: You can't get something from nothing. In the investment phase, you put in energy, but you expect to get a payout that exceeds what you put into it. This is what happens in the payoff phase of glycolysis.

So, how does it start? Glucose starts out by getting *phosphorylated* (getting a phosphate group attached to it). The enzyme that does this is called *hexokinase.* This turns it into glucose-6-phosphate. Next, there is *isomerization* (switching of atoms in a molecule) to make fructose 6-phosphate; then, another kinase acts to make fructose 1,6-bisphosphate. The goal of these three steps is to make sure the molecule gets trapped in the cell (because fructose 1,6-bisphosphate can't leave the cell). It also cleaves neatly into two equal 3-carbon sugars in the next step. This is what these three steps look like:

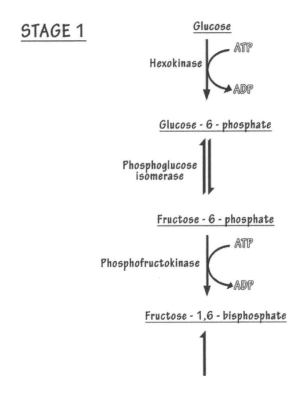

This is also the investment stage because you need to put two ATP molecules worth of energy into the process in order to allow anything at all to happen. This seems like a waste of time, but, in reality, there will be a payoff later.

The next part is stage 2. This involves the cleavage of fructose 1,6-bisphosphate into two separate but identical 3-carbon sugars. Two separate sugars are made but only the glyceraldehyde 3-phosphate goes on to the next step. The process looks like this:

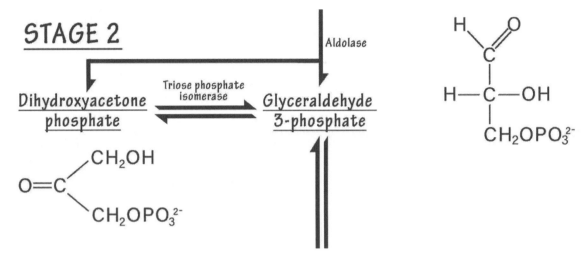

Stage 3 is the payoff stage, where ATP is finally generated. This stage ends with the making of pyruvate as the end product of glycolysis. It looks like this:

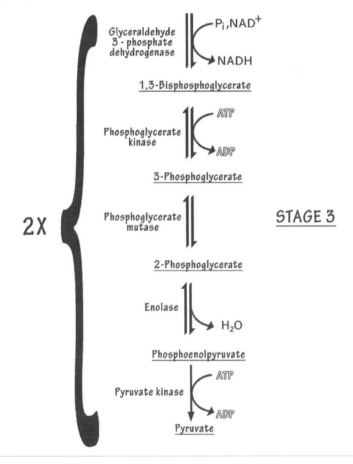

You should notice three things about this step. First, it makes two ATP molecules per molecule from the initial 3-carbon sugar (glyceraldehyde 3-phosphate). Second, there are two of these molecules per molecule of glucose, so the real yield is four ATP molecules. These are added and then the two that were used up earlier are subtracted to give a total yield of two ATP molecules per broken down glucose molecule. Lastly, you should see that a molecule of NAD+ is reduced to NADH. This molecule (two of them per molecule of glucose) goes on to help make ATP in the electron transport chain (if this is possible).

Pyruvate Moves On...

For those of us who use oxygen, glycolysis might be over, but pyruvate is still present in large numbers and there is more catabolism that can occur. From here, things move to the inner aspect of the mitochondria. There is an enzyme called *pyruvate translocase* that acts like a symporter protein system, allowing pyruvate to cross the mitochondrial membrane using the hydrogen ion as the symporter ion along with it.

Once the pyruvate gets into the mitochondria, it has a carboxyl group removed from it using an enzyme called *pyruvate dehydrogenase*. This is a large enzyme that takes NAD+ and coenzyme A as well as the pyruvate to make acetyl CoA, CO2, and NADH. It looks like this:

$$\text{Pyruvate} + \text{CoA} + \text{NAD}^+ \longrightarrow \text{acetyl CoA} + CO_2 + \text{NADH}$$

This reaction is irreversible and makes a 2-carbon molecule that enters the Krebs cycle in the mitochondria. This is the part of the cycle that completely breaks down the 2-carbon molecule into two molecules of CO2. It also makes some GTP and other molecules for the electron transport chain later.

The Krebs Cycle or Citric Acid Cycle

When you look at the Krebs cycle, you'll see that it's basically a circle that starts with oxaloacetic acid and ends with oxaloacetic acid. It circles over and over again. If the whole cycle is circular, how does anything at all happen to affect metabolism? The truth is that, even though it is a circle, molecules feed into it and leave it in a changed form. It isn't the cycle itself that changes, but the molecules that enter and leave the cycle. The cycle acts in some ways like an enzyme system that doesn't really change but helps other molecules change.

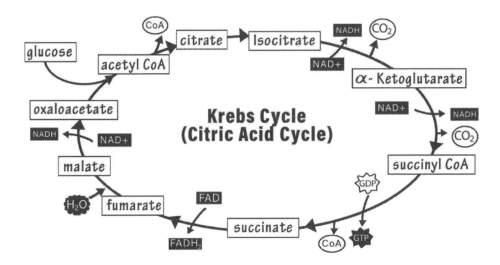

You see where the acetyl CoA is fed into the system after having come from pyruvate? The coenzyme A part (CoA) leaves immediately and the first molecule you come to is citrate, which is a 6-carbon molecule made from the two carbons you get from pyruvate and the four carbons seen in oxaloacetate. The reaction looks like this:

Oxalocetate   Acetyl CoA   Citryl CoA   Citrate

This is the start of the process that essentially cleaves off the two carbons added to the cycle at the beginning by giving off two molecules of carbon dioxide at different points in the cycle. The breakdown stops here so that at the end of the cycle, there is *oxaloacetate* left in order to begin another turn of the cycle.

If you've kept track, you started with a 6-carbon sugar of glucose. Before you even get to the Krebs cycle, this six-carbon sugar becomes two 3-carbon sugars and one carbon is dropped off of each pyruvate molecule, leaving four total carbon atoms left of the original sugar molecule. In the cycle (which runs twice for the two pyruvate molecules left), two carbons get dropped off as CO2 per cycle turn. This burns up the whole molecule.

You might also think that nothing else really happens because no ATP is made in this cycle. What good is it then besides burning up glucose? As you look at the overall outcome of the cycle, you can see that things happen even if no ATP is made directly:

$$\text{Acetyl CoA} + 3\ NAD^+ + FAD + GDP + P_i + 2\ H_2O \longrightarrow$$
$$2\ CO_2 + 3\ NADH + FADH_2 + GTP + 2\ H^+ + CoA$$

There are other energy molecules made in the cycle that will turn out to be very important later. A total of three NADH molecules are made (six total per glucose molecule) and one FADH2 molecule (two total per glucose molecule) get made in the cycle. As you'll see in a minute, each NADH molecule makes about 2.5 molecules of ATP energy and each FADH2 molecule makes about 1.5 molecules of ATP energy in the electron transport chain that happens next. If you do the math, it means that about 10 molecules of ATP get made out of each acetate molecule in the cycle (or 20 per glucose molecule).

Even though you don't see any oxygen in this cycle, the process does require oxygen. This is because you need oxygen to regenerate the NAD+ and FAD in the mitochondria. Without oxygen, these molecules would be used up and the process would quickly grind to a halt. The NAD+ needed to operate the glycolysis pathway is, however, able to be regenerated because pyruvate goes to lactate in muscle cells under no-oxygen conditions (anaerobic conditions) and this regenerates the NAD+ molecule.

The Role of the Mitochondrion in These Processes

The mitochondrion has two membranes called the *outer membrane* and the *inner membrane*. The outer membrane is the boundary of the organelle, while the inner membrane is highly folded and is where the real action is inside the structure. This is where *oxidative phosphorylation*, or the *electron transport chain*, takes place. The spaces you will need to understand about this structure are the *intermembrane space* and the *matrix*. Here's where these spaces are located:

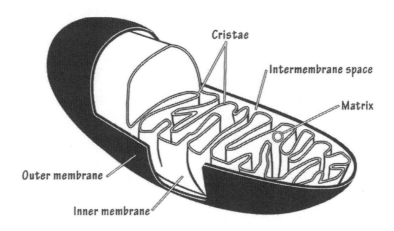

The outer membrane is very permeable and contains a *porin protein*. *Porins* are transmembrane proteins that act as channels for different molecules to pass through. The porin protein on the external mitochondrial membrane is called *VDAC*, which means *voltage-dependent anion channel*. It is helpful to the mitochondrion because it allows many organic and inorganic ions to pass into and out of the organelle.

The inner membrane is really amazing. It separates the intermembranous space and the mitochondrial matrix. What makes it so important is that it has many different embedded enzymes in it that together allow the electron transport chain to happen and for large amounts of ATP to be made. In the next section, we'll look carefully at how it all works.

Electron Transport Chain Mechanisms

The electron transport chain, or *oxidative phosphorylation*, happens across the inner mitochondrial membrane. It's probably best to see how it all works by looking at the different steps necessary to drive this process so that ATP can be made. The enzymes involved are truly "motors" in many ways because they drive a chemical reaction that does work. The end result of that work is that ATP molecules get made.

Let's look at step one, which is to create a *proton motive force*. This is a good name for it because it involves protons (hydrogen ions) that undergo some kind of movement. Finally, this movement creates a force that drives a reaction to make ATP energy.

The proton motive force is made by tearing off the hydrogen ions from the NADH and FADH2 molecules. Electrons are removed as well. The electrons get sent to different parts of the electron transport chain down an energy gradient.

The molecules accepting these electrons are called *electron acceptors*. Each time the acceptance occurs, a proton gets pumped into the intermembranous space so there is a net positive charge in this space compared to the mitochondrial matrix. This electrochemical gradient across the membrane is the *proton motive force*. This is what it looks like:

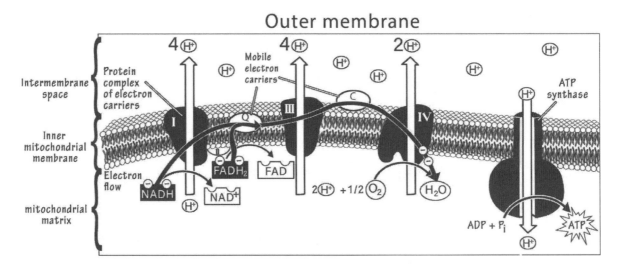

As you can see, there are four protein clusters, named *Complex I* through *Complex IV*. The electrons get passed this way:

1. NADH gets oxidized, sending electrons to Complex I. This pumps four hydrogen atoms into the intermembranous space.
2. FADH2 enters at Complex II, using coenzyme Q, which scoops up the electrons from both Complex I and Complex II, sending them to Complex III.

3. Complex III then pumps four hydrogen atoms across the membrane, using a protein called *Carrier C* in order to send the electrons to Complex IV.
4. Complex IV completes the pumping of hydrogen atoms by sending two more of these across the membrane. The remaining two hydrogen ions are added to half of an oxygen molecule to make water. This water accounts for 20 percent of the water found in your body.
5. ATP synthase uses this proton motive force to drive ATP synthase, which is a "tiny motor" that adds a phosphate group to ADP to make ATP. (Remember that this is where heat energy comes from as well because the entire process isn't terribly efficient.)

Step 2 is the *chemiosmosis* part we looked at earlier. Protons travel back through the membrane in order to balance the charges and concentration differences across the membrane. The ATP synthase enzyme is a rotary enzyme that literally rotates as part of the activity it needs to have in order to make the ATP molecule.

Step 3 is where these electrons finally lose the most energy so they reach the lowest energy level they can possibly have. They reach the oxygen molecule, combined with the protons now present in the matrix after going through the inner mitochondrial membrane. The end result is that oxygen plus hydrogen ions combine together to make water. A total of four water molecules get made for every eight protons getting through the membrane.

To Sum Things Up

The purpose of glucose and other nutrients is for the cell to use these as fuel. We didn't talk about fatty acids and their use as fuel, but these types of molecules are essentially all chemically transformed into molecules that can be funneled into the same biochemical pathways we just discussed.

The three combined pathways include anaerobic parts and aerobic parts. The first part is anaerobic, so no oxygen is necessary. It involves glycolysis and the breakdown of glucose as a 6-carbon sugar into pyruvate, a 3-carbon molecule. Two of these are made out of one glucose molecule.

The next two steps require oxygen. The Krebs cycle breaks down the entire glucose molecule, providing molecules of NADH and FADH2 that are later used to support the electron transport chain in the mitochondria.

This last part (the electron transport train) involves complexes in the inner mitochondrial membrane that take electrons from NADH and FADH2, sending them in a bucket brigade from one complex to another until enough hydrogen atoms are pumped into the intramembranous space to make the proton motive force. The final electron acceptor is oxygen, which gets reduced to make water.

This proton motive force is what drives the ATP synthase enzyme complex. As this enzyme complex is powered, it makes the ATP necessary for the entire rest of the cell to function. The whole thing makes about 32 ATP molecules per molecule of glucose burned up.

CHAPTER 22:
INTRODUCTION TO PHOTOSYNTHESIS

Photosynthesis is necessary for all living things in the world. Plants aren't the only organisms that participate in photosynthesis, but they are the only ones that have *chloroplasts* that perform this important series of reactions.

The whole goal of photosynthesis is to take light energy and carbon dioxide from the air in order to build glucose molecules from scratch. This is the ultimate in anabolism because six tiny CO2 molecules make a whole 6-carbon sugar. Chloroplasts are the organelles where all the actions happen.

Let's look first at the larger issue of photosynthesis. Later on, we will talk about how it all works biochemically. In the chloroplasts are stacked membranes called *thylakoid membranes*. These membranes are filled with pigments that absorb photons (units of light energy) in order to use the energy from the light to drive the necessary reactions needed to make glucose.

Pigments are Necessary for Photosynthesis to Happen

It's the *photopigments* in the plant that get the whole ball rolling. Their sole purpose is to absorb photons of light. Like any pigment, they absorb certain frequencies of light and reject others. This is what makes them *pigmented*, or colored. Each pigment has a set pattern of absorption, so it only absorbs those photons with a certain energy or light frequency. Red light has a long wavelength pattern while blue light (on the opposite side of the rainbow) has a short wavelength pattern.

The three main plant pigments are *chlorophyll a, chlorophyll b,* and *beta-carotene.* The color of each pigment isn't the color of the absorbed frequency of light; it is the color of the rejected, or "reflected," frequency of light. Beta-carotene is orange because it mostly rejects the orange color. The chlorophylls are more similar to what you think of when you see plants; they are green in color. There are other, less common pigments in photosynthesis called *chlorophylls c and d*, as well as the *bacteriochlorophyll* seen in bacterial organisms.

Chlorophyll has a tail made of a hydrocarbon that is so hydrophobic that it sits nicely in the thylakoid membrane. The rest of it is a globular structure called the *porphyrin ring*. It looks a lot like the heme part of hemoglobin, but instead of iron as is seen in the hemoglobin molecule, there is a molecule of *magnesium.* This is the part that absorbs light.

All photosynthetic plants have chlorophyll a in them; some will also have other pigments. Chlorophyll a is so important that the other pigments are called *accessory pigments.* The accessory pigments are used so that certain plants can gather other frequencies of light.

When any of these pigments absorbs light, it means that a photon of light energy strikes an electron, moving it to a higher energy state, or an *excited state.* In atomic terms, it bumps the electron to a higher orbital than its ground-state orbital.

Based on quantum mechanics principles you've already studied, you know that the electrons become excited only when the right wavelength of light hits them. It's like striking a pool ball with a pool stick. If you strike it on the edge of the ball or miss it altogether, nothing much will happen. If you strike it perfectly, though, the ball takes off in the direction you want it to. In the same way, the electron gets excited when the light energy strikes it perfectly. It looks like this:

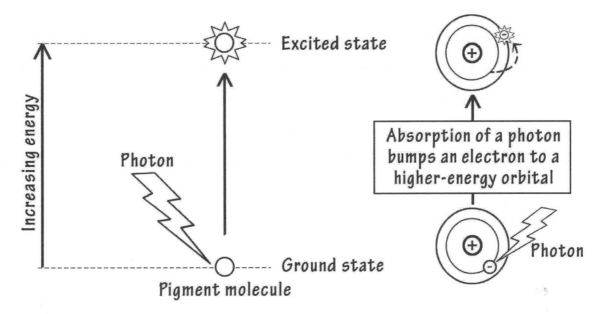

In order for the molecule to be stabilized again, it has to give off this energy to another molecule or give its electron away. This is how the light reactions in photosynthesis work.

The Overall Process

There are two main parts of photosynthesis: 1) the light reactions and 2) the dark reactions. The light reactions set up the energy necessary to run the dark reactions. The three things that happen as part of light reaction processes are:

1. Making a reducing molecule, NADPH, in the chloroplasts (which is similar to the NADH used in oxidative phosphorylation)
2. Making a proton motive force in order to make ATP energy
3. Giving off oxygen gas molecules as a byproduct of these processes

Chlorophyll is an amazing molecule. It has a central magnesium ion and absorbs light. As it does this, it pushes electrons up to a higher energy state. These excited electrons leave behind a grouping of molecules in the photosynthetic complex called the *special pair*. As electrons leave the special pair, they go to a quinone molecule and force an electrochemical gradient in the chloroplast to happen. This gradient helps to drive ATP synthesis in ways strikingly similar to what happens in the mitochondria. Take a look at an overview of this process:

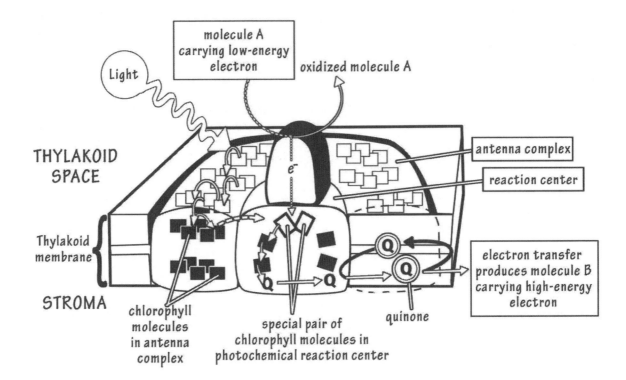

After the light reactions happen, the dark reactions happen. These are called *dark reactions* because the energy of light has already been put into the system, so the dark reactions can take place using this energy. They no longer need the immediate impact of light because the energy to run them has already been created.

To Sum Things Up

The process of *photosynthesis* is what happens in plants and other organisms that take light energy and turn it into chemical energy that can be used to drive reactions. The reactions that happen after this light energy is captured are *anabolic*. In other words, they take smaller molecules of CO2 in order to make larger molecules of sugar and complex polysaccharides.

The whole process happens in chloroplasts in eukaryotic cells. These are organelles that have pigmented molecules in them. The pigments capture *photons* of light and get energized. The energized molecules are able to transfer their energy to drive a proton motive force that will make ATP energy in a similar way to mitochondria.

In the next two chapters, we will talk about the two halves of photosynthesis: the light reactions and the dark reactions. When these two parts are coupled, energy is used to make the sugar molecules animals rely on.

Chapter 23:
Light-Dependent Reactions

The main purpose of photosynthesis is to make glucose from carbon dioxide. There are processes that can go on to make the common disaccharide sugar sucrose, which makes things sweet. Leaf starch is another macromolecule made. This starch is stored inside the chloroplast. Some starch is stored in root systems of plants (think potatoes) and others go on to become parts of fruit like seeds as well as other things we eat from plants. How plants make these nutrients is fascinating!

The light-dependent reactions involve the first three out of four total stages of photosynthesis. You can divide photosynthesis into the three main light reaction parts: 1) light absorption, 2) electron transport to make NADPH, and 3) the making of ATP energy. Next comes the dark reaction part: 4) the making of the carbohydrates (called *carbon fixation*). These are all interconnected parts necessary for the making of these carbs.

The light-dependent reactions in photosynthesis are where light energy gets transformed into chemical energy. This chemical energy drives the biochemical pathways that make large molecules out of very small ones. You can argue that the basis of our life is light and what these photosynthetic organisms can do with it. Let's look at how light gets turned into chemical energy.

Summary of the Light-Dependent Reactions

It all starts with light absorption. Light has a lot of energy in it. One mole of photons has been calculated to contain more than 50,000 calories in it. This is enough energy to make several moles of ATP out of just one mole of light energy. It is relatively efficient and very effective. Research shows that the process is about 26 percent efficient, which is somewhat less than *oxidative phosphorylation.*

Light must start the process by striking chlorophyll and other pigments in the *thylakoid membranes.* Light energy strikes these molecules and starts the process by taking water and breaking it up. The reaction looks like this:

$$2\ H_2O \xrightarrow{\text{light}} O_2 + 4\ H^+ + 4\ e^-$$

There are free hydrogen atoms and free electrons made in the process. These electrons get transferred from the chlorophyll to a primary electron acceptor, which is called Q for a molecule named *quinone*, similar to the *Coenzyme Q* molecule in animals. (Remember that Coenzyme Q is an electron acceptor in the electron transport chain.)

The Q molecule starts a chain of electron transport acceptor molecules in the thylakoid membrane that's similar to what occurs in the mitochondria. The final electron acceptor in this case is different than in animals; it isn't oxygen but is instead NADP+, which gets reduced to NADPH. All of this causes the pumping of hydrogen atoms across the thylakoid membrane and the presence of a proton motive force. This adds to the above chemical reaction, so it now looks like this:

$$2\ H_2O + 2\ NADP^+ \xrightarrow{\text{light}} 2\ H^+ + 2\ NADPH + O_2$$

Bacteria often have a variation of this that doesn't involve NADPH. Instead, they use hydrogen gas or hydrogen sulfide and the electrons in these molecules to make NADH but not NADPH.

This image shows the specifics of photosynthesis and the electron transport chain in the chloroplast:

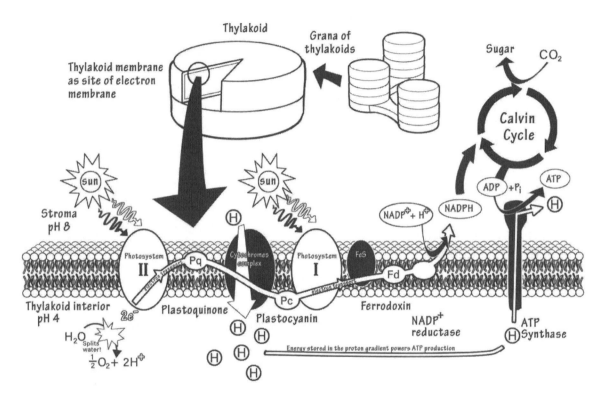

The end result is that ATP gets made in the same way that it happens in mitochondria. There is a proton motive force made in the thylakoids. These protons get moved from the interior of the thylakoid to the *stroma* around them. There is a complex called the *FOF1 complex* that is coupled to the synthesis of ATP as it uses ADP and phosphate to create the energy-producing ATP molecule. This ends the light-dependent reactions.

In More Detail: There are Two Photosystems Involved

The whole photosynthetic process in the light-dependent reaction side of things involves two different photosystems that are linked together. If the whole thing was a factory instead of a complex of molecules, it might look like this:

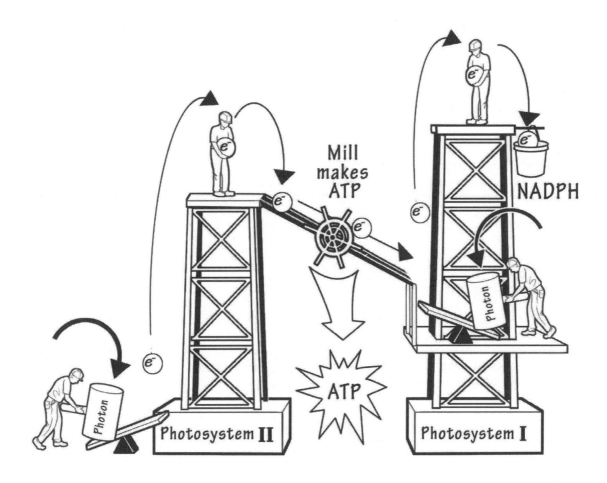

In the image, you have workers in Photosystem I that take a photon plus a molecule of NADP+. The photon bumps up the energy of the electron in order to make NADPH. In Photosystem II, you have another electron that gets bumped up in energy. This energy is used to generate ATP by oxidizing water to make oxygen, electrons, and hydrogen atoms. This is how light gets chemically harvested for use to make energy-making molecules.

In the chloroplast, the process is similar but looks like this:

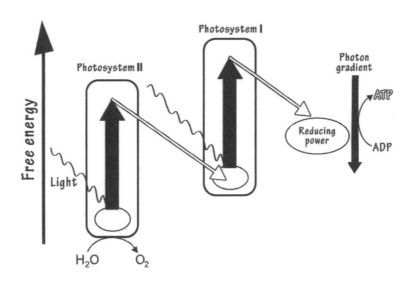

The left-hand side is where Photosystem II is located. The chlorophyll captures the photon energy and uses the energy to create both NADPH and ATP. The making of these two molecules is what the light reactions are all about. The right-hand side of the system is Photosystem I, set at a higher energy level. This also uses photon energy to drive the process of making ATP energy.

A great deal of energy gets stored this way — about 4×10^{17} Kilojoules (or about 10^{17} kilocalories) of pure energy are made and stored by all the plants on Earth every year. This energy makes 10^{10} tons of carbs and other organic molecules per year.

That's a lot of carbs! In fact, it is estimated that if ordinary sugar cubes were made by all photosynthetic organisms on the planet, the sugar cubes stacked on Earth would reach Pluto. Instead, a variety of molecules gets made through these processes. We eat them and transform them into all of the structural and functional molecules in our cells.

What's in These Photosystems?

Instead of the complexes in mitochondria, photosynthesis has two major complexes, or *photosystems.* Each photosystem has many different proteins in a cluster. All of these are in the thylakoid membrane. The process starts with pigments called *antenna-containing pigments*. These are clustered around a reaction center that has two chlorophyll molecules and a bunch of proteins in it.

The antennas are named because they are like radio antennas that pick up radio signals on a portable radio. They do not actually look like antennas. Like a radio, they are tuned to a certain frequency of waves (except that these are light waves and not sound waves). Light is captured by the antennas by *LHCs*, or *light-harvesting complexes;* these complexes essentially funnel the light to the pair of chlorophyll molecules, sitting next to each other in what's called the *reaction center.*

This image looks at the collection of light in a light-harvesting complex. Light hits the *antenna chlorophylls,* where it gets funneled to the pair of chlorophylls in the reaction center. These bump up the electrons to the Q, or quinone, molecule. The hydrogen atoms or protons get left behind, which is how this *proton motive force* gets created in the thylakoid membrane using Photosystem II. The wavelength of light used to drive this reaction is set at 680 nanometers.

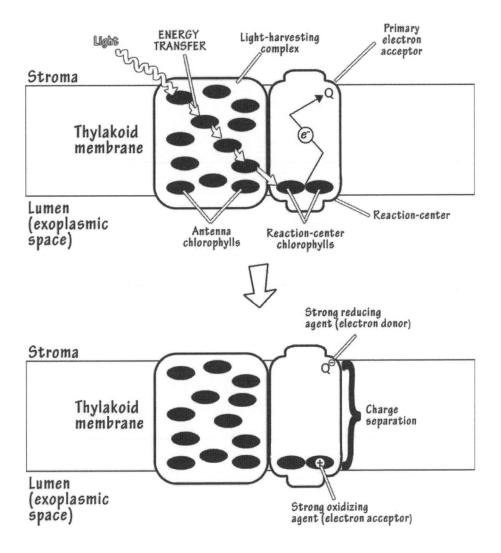

Chlorophyll is a weak reducer, which means it can accept electrons from another molecule but doesn't give it away very easily. BUT … if you give it an excited electron, the whole process is much easier. Excited electrons are in an outer orbital so that they are more readily given away (and accepted by the Q molecule). It is much easier by having the electron energized by light.

The Q molecule is able to transport the electron from the excited chlorophylls in the reaction center. The protons do not go with the electron, so the gradient of protons on one side of the membrane occurs. Q goes on to give electrons to other electron acceptors. This goes on until NAPD+ accepts the electron as the *final electron acceptors.*

Excited chlorophylls are very powerful. They can split covalent bonds and facilitate the formation of others in an attempt to get back to its original unexcited state. If "P" is the unexcited chlorophyll and P+ is the excited chlorophyll, the whole thing looks like this:

$$2\ H_2O + 4\ P^+ \longrightarrow 4\ H^+ + O_2 + 4\ P$$

What is the purpose of the extra pigments other than the main chlorophyll a molecule? It turns out that, even at noontime on a tropical sunny day, only one photon per second can be captured by chlorophyll itself. With the extra pigments involved, more wavelengths of light can be captured, and more photosynthesis can occur with a greater degree of efficiency.

Photosystem I catches wavelengths of light set at 700 nanometers or below, which is different from the 680 nanometers of light captured from Photosystem II. Electrons flow first through Photosystem II and then through Photosystem I. This is what they look like together on the membrane:

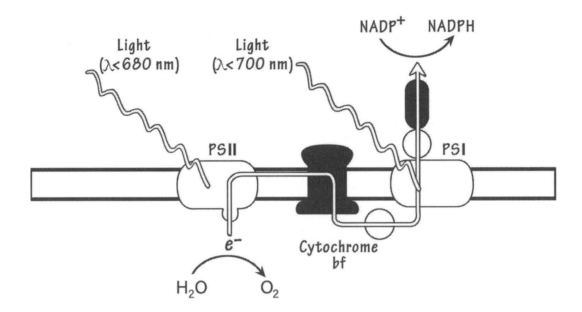

As you can see, NADPH gets made only when the process goes through Photosystem I. The energy collected here goes on to make NADPH directly.

To Sum Things Up

The light reactions of *photosynthesis* occur in the *thylakoid membranes* and involve the capture of light energy. This light energy is captured by pigments in two photosystems. One photosystem uses the electron excited by the addition of light energy in order to cause two chlorophyll molecules in the reaction center to split water. This is where oxygen gets made, hydrogen atoms are used to create a proton motive force, and electrons start down an electron transport chain.

At the end of the electron transport chain, NADP+ accepts the electrons and gets turned into NADPH. We'll talk next about why you would need so much of this molecule to proceed with photosynthesis. Another endpoint of the light-dependent reactions is that ATP energy gets made. This ATP energy is used by the cell for several things. Even in plant cells, ATP is the main energy currency these cells need.

CHAPTER 24:
CALVIN CYCLE

The Calvin cycle is the main activity that happens in photosynthesis to fix carbon into making the glucose molecule. This is the *dark reaction segment* of photosynthesis. It doesn't really depend on it being dark to have these reactions happen; the point of it being called "dark reactions" is that the energy to run these reactions has already been created and its functions can occur even in the dark.

The Calvin cycle happens in the interior, or *stroma*, of the chloroplast, which is where starch granules are kept after being put together by the many glucose molecules made by the carbon fixation process. Here's what the Calvin cycle looks like:

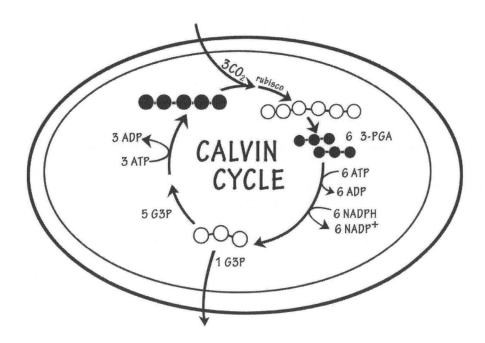

The circles in the image are carbon atoms. You can see where six carbon dioxide molecules are added to the cycle to a mysterious 5-carbon molecule called *ribulose-1,5-bisphosphate* (also called *RuBP*). It's one of those molecules that doesn't really go towards making anything itself. Like the Krebs cycle, there are molecules (RuBP mainly) that just keep getting recycled as other molecules get added and subtracted from the cycle.

A lot of ATP and NADPH are also a part of this cycle. These are used for energy to drive the reactions necessary to put the CO2 molecules together. The enzyme that starts the whole thing is called *rubisco*. Rubisco binds CO2 to the RuBP molecule one at a time to make a 6-carbon molecule. This 6-carbon molecule gets split into two 3-carbon molecules. This process of making CO2 into an organic molecule is what we call *carbon fixation.*

The 3-carbon molecules that get made from this cycle are called *G3P,* which stands for *glyceraldehyde-3-phosphate.* These are the molecules that leave the cycle, not glucose. They still have to go on later to make a glucose molecule. It takes three turns of this cycle to make one molecule of G3P, so it takes six cycle turns in total to make enough fixed carbon to make glucose.

Only one out of six G3P molecules made in the cycle actually leaves the cycle to make sugar. The rest of the cycle makes five G3P molecules that are called *leftovers*. These leftovers use ATP energy to keep feeding into the cycle so that even more CO2 can be added to restart the cycle again.

This is what the RuBP, or ribulose 1,5-bisphosphate, molecule looks like. It has five carbons and two phosphate molecules on either end of the sugar:

CO2 from the air enters the cycle and mixes with the 5-carbon RuBP. Together, these get acted on by the Rubisco enzyme. This process leaves a combined molecule that has six carbon atoms. This 6-carbon sugar is very unstable, so it breaks down immediately into the two 3-carbon G3P molecules.

The Rubisco enzyme complex is found in large amounts in plant cells. It is thought to be the most abundant protein on earth. Of leaf proteins, Rubisco makes up about 20 to 25 percent of the total by weight. It does not require ATP energy by itself, so in dead and decomposing plants, it keeps working in order to facilitate the decomposition of plant material.

Biochemically speaking, the Calvin cycle looks like this next image. It really isn't too important to memorize the chemical structures. Instead, look at the carbon atoms and how the cycle starts with one 5-carbon molecule, which then goes on to make two 3-carbon molecules after CO2 is added. Some of these stay to get recycled, while some go on to make sugars.

This is a brief summary of the recycling of the five 3-carbon G3P molecules that happens to feed the Calvin cycle again:

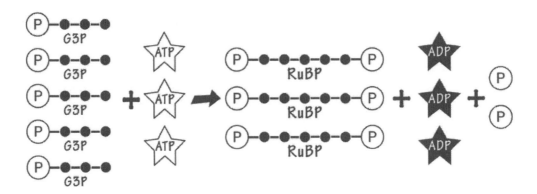

G3P (not glucose) is the end product of the Calvin cycle and of photosynthesis itself. This is a versatile molecule that can get rearranged in different ways to make glucose and starch in the chloroplasts. The process of making glucose from scratch this way is called *gluconeogenesis.* Gluconeogenesis is similar in plants and animals; the difference is the source of the carbon atoms that start the process of making glucose. Basically, it looks like this:

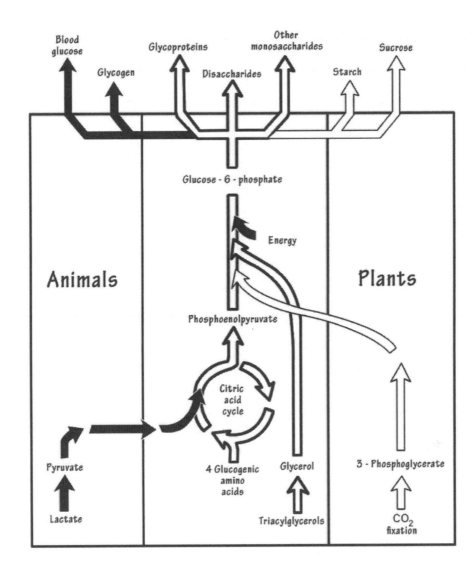

Gluconeogenesis isn't the exact "opposite" of glycolysis because it involves different enzymes and intermediaries. One main thing they do have in common, though, is that G3P is part of both biochemical pathways. G3P can also make other monosaccharides and be used by itself as a nutrient, where it is burned up in the glycolysis pathway as part of cell respiration in the plant cell.

Take a look at the glycolysis pathway again to see how this works. G3P is one of the early intermediates for the glycolysis pathway. This means that plants do not have to make glucose completely in order to use G3P instead part of the way through the glycolysis pathway. The end result is that the plant cell can make ATP energy directly out of the G3P molecule itself.

To Sum Things Up

The dark reactions of photosynthesis mainly involve *carbon fixation* and the *Calvin cycle.* This cycle keeps turning, adding one carbon dioxide molecule at a time in order to make carbon as an inorganic molecule (in CO2) into an organic carbon-based molecule.

The end result of the Calvin cycle and photosynthesis isn't glucose. The end result is really *glyceraldehyde-3-phosphate*, or *G3P*. Five of the six molecules of G3P made are recycled as part of the Calvin cycle.

One of these G3P molecules leaves the cycle in order to make glucose, starch, and all other molecules on Earth (at least in part). Some of the G3P molecules enter the glycolysis pathway partway through the process so it can be burned up again in cell respiration in the mitochondria in plant cells (the same way it happens in animal cells).

SECTION SEVEN:
CELL MOTILITY AND DIGESTION

This section talks about the main proteins involved in determining the cell's structure. These proteins form the cell's *cytoskeleton*. It turns out that these are the same proteins used to help certain cells have motility, or movement. As you'll see, cells can move in different ways.

Some of these proteins also participate in attachments between adjacent cells, using anchoring proteins that attach the cells (not from their cell membranes only but using the cytoskeleton to anchor the cells together). There are different attachments possible between the cells.

Finally, we'll discuss *cell digestion,* which is the same thing as cell respiration. Cell digestion does not produce energy for the cell but only takes larger things inside or coming from outside the cell and breaks them down. Some of these breakdown products get recycled, while others get excreted from the cell surface through *exocytosis.*

CHAPTER 25:
CYTOSKELETON

Just like we humans have a skeleton that gives us our shape, so we aren't just blobs, the cell has a certain cytoskeleton to give it some structure (even though many of them still look like blobs anyway). The cytoskeleton of the cell is made from different proteins linked together in a long chain.

Besides giving a cell its structure, the cytoskeleton allows cells to move. Cells can move by streaming their cytoplasm, causing the cell to creep across a surface. They can also have *flagella* or *cilia* that move the cell using proper machinery designed just for this purpose.

There are cytoskeletal proteins that help organelles move within the cell itself (from one place to another) or that keep organelles fixed in one place. Things like *endocytosis* and *cell signaling pathways* rely on the cytoskeleton for their activity. When cells divide, parts of the cytoskeleton are necessary for the chromosomes to separate and for the daughter cells to divide completely.

Lastly, the cytoskeleton can help anchor two neighboring cells to one another. The cytoskeleton will have proteins that stick to the cell membrane so that when two cells attach to one another, the attachment is made firmer and more structurally sound by these cytoskeletal proteins.

Proteins in the Cytoskeleton

All cells (*prokaryotes* and *eukaryotes*) have some type of protein associated with them. The different types of cytoskeletal components can be defined by the type of protein they are made of. As they come together, they form a meshwork pattern inside the cell. This is what it might look like inside the cell:

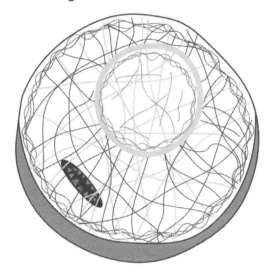

cytoskeleton

There are three different types of cytoskeleton *filaments* or *fibers*. These are called *microtubules, microfilaments,* and *intermediate filaments.* Why have three separate kinds of fibers? It's because each of these fiber types has a unique structure and function based on what it's used for. Let's briefly outline these:

- **Microtubules** — These are thick, hollow rods made of a protein subunit called *tubulin*. They are the most important fibers for cell movement both inside the cell and for the cell itself as it moves through its environment. All eukaryotic cells have microtubules that are about 25 nanometers in diameter.
- **Microfilaments** — These are very thin fibers only 8 nanometers in diameter. These are usually made from an actin protein fiber. You may have heard of *actin* and *myosin* as being the major fibers used in muscle contraction in humans, but actin has many more functions than this. These fibers help organelles move inside the cell.
- **Intermediate filaments** — These are middle-sized fibers with a diameter of about 10 nanometers. Several different proteins make up these fibers. They are thicker than microfilaments, so they are particularly good as supporting filaments for the cell. They hold the organelles and even the other fibers in place. Different intermediate filaments are seen in different cell types.

Motor Proteins in the Cell

Motor proteins are those that participate in cell movement but aren't a direct part of the cytoskeleton. Without these motor proteins, the cytoskeleton would be much less functional. These proteins actually do work inside the cell. What this means is that rather than just getting made and sitting there, motor proteins need ATP energy for them to drive movement within the cell.

The three kinds of motor proteins are:

- **Kinesins** — These are the main microtubule-moving proteins in the cell. They move microtubules and anything else the microtubules are attached to. They will, for example, help move organelles toward the plasma membrane of the cell.
- **Dyneins** — These are like kinesins and also act on the microtubules. Instead of moving things away from the nucleus of the cell, they move things toward the nucleus. They work by sliding different microtubules next to each other and are the main proteins seen in the movement of cells that have cilia or flagella.
- **Myosins** — These are proteins that make a thicker filament than actin. They are combined in a unique way with actin filaments in muscle cells. The actin filaments slide along the myosin filaments so that the muscle fiber can become shorter as it contracts the cell (and your entire muscle if you have lot of these fibers). They also help with *cytokinesis*, which is the separation of two daughter cells after they get made in cell reproduction, and with both *endocytosis* and *exocytosis.*

What is Cytoplasmic Streaming?

The cytoskeleton is necessary for what's called *cytoplasmic streaming.* This is exactly what it sounds like — the cytoplasmic contents can flow like a stream in the cell. This also called *cyclosis* and is what helps cell contents circulate in the cell. Anything that needs to get from one place to another in the cell goes via *cytoplasmic streaming.*

The process happens as different microfilaments contract or move against one another in complicated ways that will allow the intracellular structures to be directed in various ways. It is also how eukaryotes like amoebae move. They use cytoplasmic streaming to flow in one direction in order to allow the cell to flow like a blob in a specific way. This is how an amoeba does this:

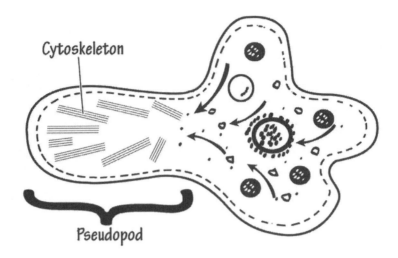

More on Microfilaments

Like all of these fibers, *microfilaments* are made from monomer proteins, or *subunits*. With microfilaments, the monomer protein is *actin.* When these subunits come together, the microfilament looks like this:

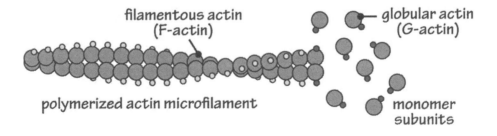

There are different ways that microfilaments work. Sometimes, the end grows and pushes against a cell membrane or other structure as it is growing lengthwise. It can also walk along other fibers so that it can direct movement in ways similar to what happens with muscle contraction. There are two types: *G-actin* and *F-actin*. The microfilament is made by intertwining two fibers — one made of G-actin and the other made from F-actin.

Microfilaments have many jobs, including cell movement, muscle contraction, cell transport within the cell, cell shape maintenance, cytoplasmic streaming, and cytokinesis. Because of their importance, all cells must have functional microfilaments in order to survive. There are a few rare diseases where the microfilaments don't work but having bad microfilaments probably isn't compatible with life.

Intermediate Filaments

Intermediate filaments are a little more complex than microfilaments. This is because there is more than one type of intermediate filament. These are stronger than microfilaments, so they are used mainly for structure. They have the ability to anchor the organelles in space and act a lot like the tent poles in your tent. They spread out in a scaffold in order to allow the cell to have some shape. These are the main structures used to anchor cells that are stuck to one another.

The structure of an *intermediate filament* is different from microfilaments. The monomer is a long protein that combines with identical long proteins. Four of these intertwined with each other make a *tetramer* of proteins. A bunch of tetramers makes a short length of filament called a *ULF*, or *unit length filament*, about 60 nanometers long. These will then connect with one another to make the whole filament of whatever length is necessary. Here is how this works:

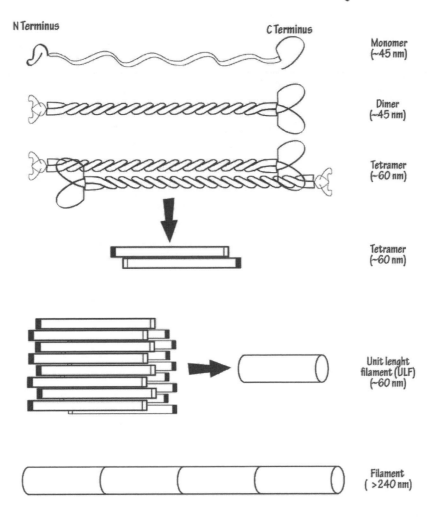

In the epithelial cells that line your GI tract or that are the surface of your skin, the main protein in the intermediate filaments is *keratin*. This is a strong protein subunit that provides the structure these cells need. They help the cells stay strong in spite of the stress they must always endure. Some biologists believe these fibers keep the cell from collapsing and dying under stress.

You should think of intermediate filaments as "scaffolding" structures. They help cells communicate with one another by maintaining the connection between them. If your intermediate filaments aren't working, you might have a disease of premature aging. Related diseases linked to some failure of intermediate filaments in children include muscular dystrophy and Alexander disease (a disease where the nervous system breaks down from intermediate filament failure).

Besides keratin proteins, intermediate filaments can be made from *vimentin proteins* (in connective tissue cells), *neurofilament proteins* (in nerve cells), *lamin* (providing structure to the nuclear envelope), and *desmin* (in muscle cells).

Microtubules

Microtubules are thick and are essentially "tubes," which explains their name. They are made of two types of tubulin proteins called *alpha-tubulin* and *beta-tubulin.* When they are put together to make a single microtubule, the structure looks like this:

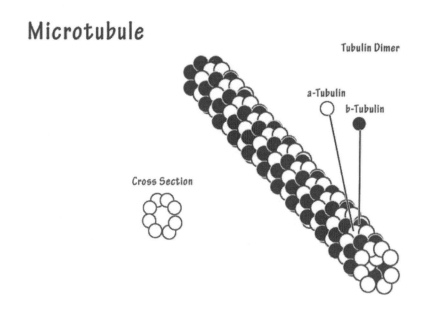

The inner tube lumen is about 15 nanometers in diameter while the outer microtubular diameter is close to 23 nanometers. These tubes require more energy than the other filaments in order to be constructed. It takes GTP energy and not ATP energy to do this. They are made in interesting structures in the cell called *centrosomes.* A centrosome looks like this:

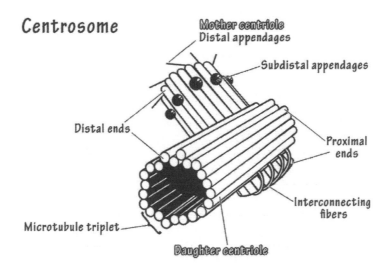

Centrosomes are made from microtubules themselves but just short segments of them. There are two structures in each centrosome called the *mother centriole* and the *daughter centriole.* These are arranged at 90-degree angles from one another. Each centriole appears as star-shaped when seen on cross-section and is made from nine triplet sets of microtubules. This means that it takes 27 (3 x 9) microtubules to build a centriole.

Centrosomes are crucial to cell division and are necessary in some organisms for the proper separation of the chromosomes whenever a cell divides. Interestingly, they don't get copied twice in the dividing cell, so each daughter cell gets just one centriole. This becomes the "mother centriole." It must make its own daughter centriole in order to have a matched set. Not every cell type has these, but certain eukaryotes always do. (Plants and fungi do not have them.) The centrosome is called the *microtubule organizing center*, or *MTOC.*

Microtubules also make up the flagella and cilia of eukaryotic cells. Here, the arrangement is different. The microtubules have a 9 + 2 arrangement pattern, which involves nine doublets plus a pair of microtubules in the middle. They use *dynein* as the main protein to connect the microtubules. The dynein acts like a motor to drive the microtubule activity. This is what the flagella arrangement looks like:

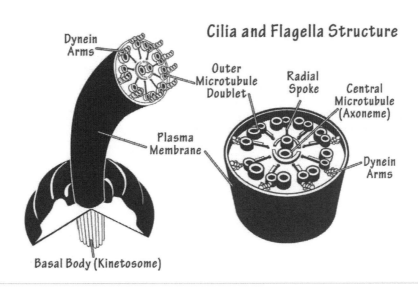

See from the image where the *dynein arms* connect the *doublets*? These proteins allow the microtubules to slide past one another in ways that don't restrict movement of the cilium or flagellum.

More on the Motor Proteins

The motor proteins are all different but are clustered together as "motor" proteins because they engage in some type of movement. The cytoskeletal proteins by themselves don't move — it's the motor proteins that help them move. Each type of filament has its own set of motor proteins. Intermediate filaments don't have these proteins; only microfilaments and microtubules have them.

The actin motors are called *myosins*. The word is plural because there is more than one type of myosin a cell can have. For example, *Myosin II* is the type of motor protein used in muscle cells in order to facilitate their contraction. Myosin is a long and skinny molecule that has two heavy chains and two light ones. The head has an actin binding site and an ATP binding site. These are used to attach these two molecules so that actin and myosin can walk along each other as a muscle contracts. This is what the myosin molecule looks like:

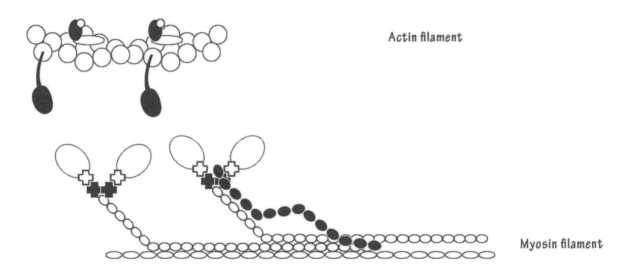

Actin filament

Myosin filament

The globular heads of myosin bind to ATP and allow it to hydrolyze. This is where the energy comes from to allow the two filaments to walk along each other. Myosin II also aids in cell division — even in non-muscle fibers.

There are other types of myosin that do other things as well, usually in non-muscle cells. Another example is *Myosin V,* which helps organelles and vesicles move in the cell. *Myosin XI* helps in cytoplasmic streaming. Again, actin can't do much by itself, so it needs myosin to allow for movement.

Microtubule motors are more complicated. There are several of these types of motor proteins. One of these is called the *kinesin class* of proteins. The movement of microtubules when kinesin is used is called *anterograde*, meaning "forward" movement. Using kinesins, chromosomes can separate in the process of cell division and organelles like Golgi bodies and mitochondria can travel from place to place within the cell.

Dynein proteins help the same microtubules but allow them to move in a *retrograde*, or "backward," direction. These are large complexes that have at least two heavy chains and a number of light chains of polypeptides. For the record, a "heavy" chain is thicker than a "light" chain. The direction of motion of the microtubules tends to be toward the nucleus, where the tubules are made by the *centrosomes*. There are dyneins that mainly work to move cilia and flagella, and dyneins that help move things within the cytoplasm itself.

Cell-to-Cell Interactions Using Cell Adhesion Molecules

Cells need to attach to each other; they also sometimes need to attach to the protein structures around them. Some connections between cells are permanent, while others are temporary. Cells use *adhesion proteins* in order to connect to other things.

If you think about it, cells of the same type should stick together in specific ways. If this didn't happen, organs like your liver or kidneys would fall apart. There is also no reason why your kidney cells should stick to your liver cells; fortunately, the body can tell the difference between these cells to ensure the right kind of attachment.

The different types of cell adhesion molecules are called *selectins, integrins, cadherins,* and *immunoglobulins.* You've probably heard of immunoglobulins, which are your basic antibodies. Antibodies are made by your immune system so that immune cells can specifically attach to other cells and to pathogens (like bacteria). These are made by certain immune cells in order help kill pathogens or infected cells.

Selectins are molecules used for temporary or transient attachments between white blood cells and the cells that line the blood vessels, called *endothelial cells.* There are three types: *L-selectin, P-selectin,* and *E-selectin.* The L-selectin is found on white blood cells called *leukocytes*. E-selectin is found on the endothelial cells. P-selectin is found on platelets.

These adhesion molecules attach to carbohydrates on other cells, allowing white blood cells to slip between endothelial cells when you get an infection. Infections in the tissues are treated partly by having white blood cells exit the blood vessels, traveling to the site of the infection in order to kill the infectious organisms. After the selectins do their job, other cell adhesion molecules called *ICAMs*, or *intercellular adhesion molecules*, help further the process so that the white blood cell can stick more firmly to the blood vessel, which finalizes the steps needed for these white blood cells to pass into the tissues.

This whole process is called *diapedesis* of white blood cells. It takes a lot of molecules to achieve this process, shown in this image:

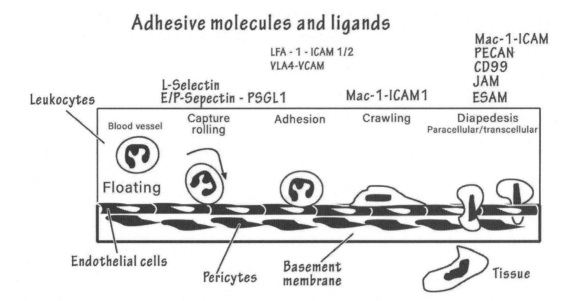

Adhesive molecules and ligands

Another cell adhesion molecule is called the *cadherin class* of molecules. These are active in embryo development and help form stable connections between similar cells in tissue. There are different types of cadherin molecules for the various types of cells and tissues.

Stable cell-to-cell connections also involve the cadherin molecules. These are used to join to cells of the same type. There are two important cell connections you should know about.

Tight Junctions

Tight junctions use cadherins to attach two cells together. These are important in epithelial cells that line the gut, for example. The tight junctions make sure that no molecules or water get between the cells of the gut lining without getting through the cells themselves. The goal is to control the flow of all molecules through these types of barriers in the body.

This is an image of a tight junction:

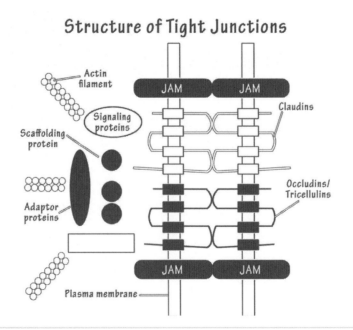

Structure of Tight Junctions

In the image, you can see the transmembrane proteins called *claudins, occludins,* and *tricellulins.* These help the cells adhere tightly to one another. There are adaptor proteins that stick to actin filaments, which anchor the membrane and the cell in ways that are firmer than if the two membranes stuck together without having the cytoskeleton involved.

Fun Factoid: *Leaky gut syndrome happens when there is a breakdown of the tight junctions in the gut lining. This is bad for you because it allows leakage into the tissues around the GI tract, causing inflammation and other diseases. Your gut can be leaky from food intolerances, malabsorption of nutrients, autoimmune diseases, and inflammation.*

Gap Junctions

Gap junctions are different from tight junctions because they actually connect two cells together in order to allow a freer exchange of messages from one cell to the other. There are open channels created between adjacent cells so that small ions and molecules can get through the membranes. They can also connect two cells together electrically. The cells of your heart and smooth muscle often communicate this way so that an electrical signal can get passed from cell to cell.

Some signaling molecules can pass through these cell membranes, too. These include *calcium ions* and *cyclic AMP*. The end result is that a signal received by one cell can affect the behavior of a nearby cell. This helps coordinate activities in a group of cells at the same time.

Special proteins called *connexins* are located in the cell membranes where gap junctions are located. These help to form tubes or tunnels between two cells. The cells are separated by a *gap*, which is why these are called "gap junctions." This is what they look like:

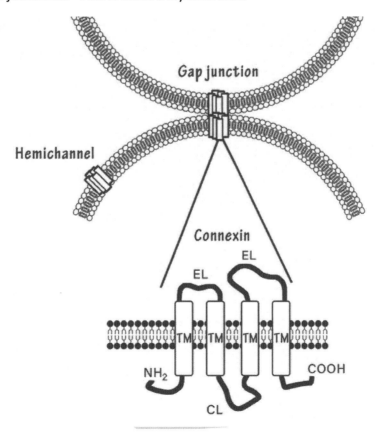

As you can see, there is a *connexin protein* on one cell membrane. This sticks out and connects to the connexin protein on an adjacent cell. This is where the channel forms between the two cells.

Desmosomes

Desmosomes are the best example of how the cytoskeleton is used to anchor two cells together. Desmosomes are also called *macula adherens* because they are one of the main ways cells adhere to one another. A *hemidesmosome* is the same thing, but attaches the cell to a protein structure, like the *basement membrane* that sits on the bottom of a layer of epithelial cells in all epithelial tissues. Desmosomes are particularly strong areas of adhesion between two cells. This makes them good options for cells under a great deal of physical or mechanical stress.

Desmosomes make use of intermediate filaments. They also make use of cadherin proteins and linker proteins. These together make a structure called a *DIFC*, or *desmosome-intermediate filament complexes*. There are three parts to these DIFCs: an *outer dense plaque (ODP)*, an *extracellular core region*, and the *inner dense plaque*. Let's see what these look like:

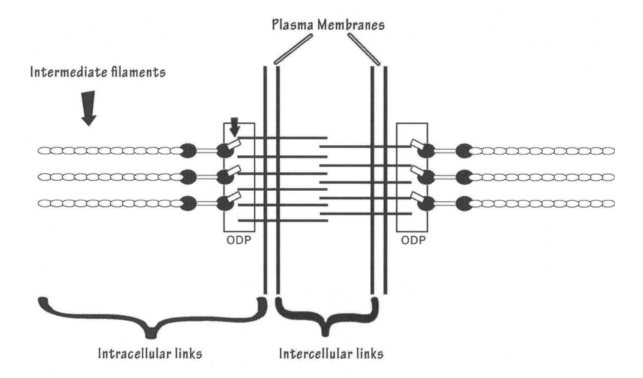

The desmosome cadherins are between the cells (in the space between them). Two of these proteins are called *desmocollin* and *desmoglein.* They attach through the membrane to the *ODP.* The outer dense plaque has two proteins: *plakoglobin* and *plakophilin.* These are the linker proteins that attach the desmosome to the cell's intermediate filaments. There are two liner sections, with an inner one labeled *desmoplakin,* which is another connector protein. See how strong this connection must be after attaching to the cytoskeleton?

Fun Factoid: *There are a lot of diseases where the desmosomes are messed up. These include certain heart diseases and the different blistering diseases of the skin. Pemphigus vulgaris is one of these; it is an autoimmune disease that attacks the desmoglein protein. The major effect is that skin cells* *separate, forming large blisters in the skin. Bullous impetigo is a Staph infection of the skin where a toxin is released by the organism that damages the desmoglein 1 protein, causing skin blistering. Still, other blistering diseases come from an inability of the intermediate filaments to attach to the desmosome.*

To Sum Things Up

The *cytoskeleton* is very important to the structure and function of the cell. Some act like the poles of a tent, holding the structure up from the inside. There are three types of cytoskeletal fibers: *intermediate filaments, microfilaments,* and *microtubules.* They look different and have different functions.

Intermediate filaments are mainly used for structure. They are attached to *desmosomes* in order to allow two cells to connect with one another in a solid way. *Microfilaments* are made from *actin* and are thin. They are used both for structure and movement within the cell. *Microtubules* are mainly filaments used in movement. They allow for movement in the cell itself as well as of the cell in the environment by operating the cilia and flagella of some cells.

There are motor proteins that help movement in the cell; the cytoskeleton doesn't do this by itself. In the next section, we will talk about cell movement and how the cytoskeleton and motor proteins come together to allow for movement in the cell and outside of it.

CHAPTER 26:
HOW CELLS MOVE

Now that you understand the cytoskeleton and know that it is important for cell movement, let's talk more specifically about how this amazing process works. We will first talk about *cytokinesis* or how internal movement can happen in a cell. Then, we will talk about how cells move in their environment (as a whole cell).

Cytokinesis: Actin and Myosin Do Their Thing

We've talked about how muscle cells contract using actin and myosin. As you've learned, though, non-muscle cells use actin and myosin to perform other functions in the cell. One of these is how actin and myosin get together to help separate a dividing cell into two daughter cells.

The myosins have been studied a lot so far: These are Myosin I, Myosin II, and Myosin V. Myosin II is found in all cells and uses ATP energy to help microfilaments made of actin achieve movement. The combination of actin and myosin in these areas is called *contractile bundles.* They are found near the plasma membrane.

There are other bundles of actin in the core of things like microvilli found on the cells of your intestinal lining. These are linked instead to Myosin I fibers. The job of all of these bundles is not so much to move the cell but to contract parts of the cell.

These are often found in what's called a *circumferential belt.* This acts like a stiff arch that tightens the cell membrane in certain areas. They are attached to the adherens junction areas of the cell and help to protect the cell from traumatic stress. They also help keep the cell at its optimal shape. This is how this belt-like structure holds a cell shape:

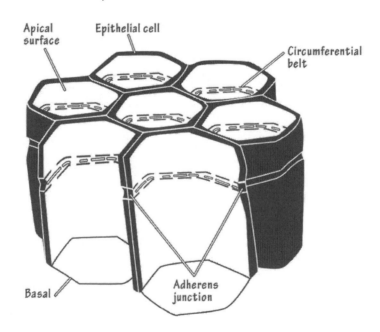

While these areas are helpful in resisting changes in cell shape, they do not do much in the way of causing any actual movement.

Actin and Myosin II help in cytokinesis. This happens in the dividing cell. The actin and Myosin II bundles attach along the *equator* of the dividing cell so that the cell can divide into two halves. They form a *contractile ring* or "purse string" that cinches the cell into two pieces. The ring shrinks until the cell is pinched into two. This is what it looks like:

Cytokinesis in Animal Cells

- Microfilaments assemble around cell middle

- Actin & myosin microfilaments form a ring around cell

- ATP (energy) used to squeeze cell in half

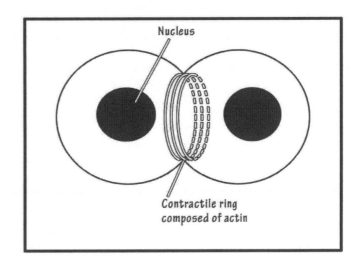

Myosins also help some vesicles in the cell move around. Vesicles don't just diffuse around the cell randomly. They are on a mission, so they have specific places to go. Myosin/actin contractile fibers help these vesicles go exactly where they are needed in ways we don't completely understand yet. It is believed that other myosins (probably Myosin I and Myosin V) are involved in this process somehow, which probably works like other actin/myosin contractile processes. Microtubules also help in some vesicle transport within the cell.

Myosin also helps in *cytoplasmic streaming*, or the general flow of cytoplasm from one place to another. Cytoplasm can actually flow easily and quickly around the cell. In one study on algae, they found that the cytoplasm circled the cell in a loop at a rate of 4.5 millimeters per minute. That's pretty fast for a tiny cell. This may be how some cells circulate their nutrients without using specific actin/myosin pathways.

This image shows you what cytoplasmic streaming probably looks like and how the same process can help a cell like an amoeba reach out and move in its environment:

Cytoplasmic Streaming

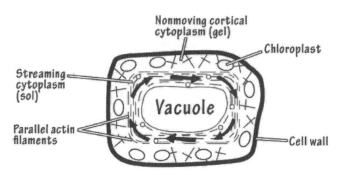

Pseudopod Movement

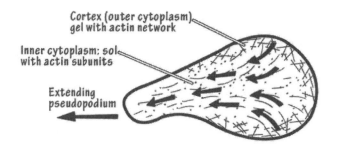

Movement of Cells in the Environment

The picture above shows what an amoeba looks like. These organisms are single-celled eukaryotes that move across solid objects by sending out *pseudopodia* ("false feet"). These are areas where the cell expands in one direction — essentially streaming itself in one way only so that it travels in that direction, dragging the rest of the cell along with it. This generates force that moves the cell.

Other cells besides amoebae can do this. *Fibroblasts* are the cells in your body that travel to a wound in order to make scar tissue. Like the amoebae, they have a specific place to go, and it wouldn't make sense for them to randomly end up there without a real sense of direction.

The whole process happens by lengthening the actin cytoskeleton (microfilaments) in the direction the cell needs to go. These microfilaments get assembled into bundles and essentially grow outward to make the cell extend in the desired direction. By growing in one direction only, the cell can be forced to travel in that direction. Outside the cell, there are sticky adhesion molecules that attach along the way to keep the cell from backtracking. By pushing forward, attaching, pushing further still, and attaching to another place, the cell can move little by little. In fact, it's a lot like an inchworm.

One of the cool parts about amoebae and similar moving cells is that it really does a "head" end and a "tail" end. The head end is the leading end where the cell extends itself. In the amoeba, these extensions, or pseudopodia, are able to sense the direction the cell "wants" to go and extends outward at this point. As you can imagine, there are external cell signals, or *cell signaling*, that control where a cell is supposed to go.

At the leading end of the *pseudopod*, there are a lot of *G protein-coupled receptors* that must bind to something that stimulates calcium release at that end. Calcium plus ATP energy gives rise to more actin getting made in order to push the process forward. In the lagging end, the myosin bundles break down, possibly because the calcium and ATP energy levels are low in that part of the cell.

Flagella and Cilia Use Microtubules

Cilia are short — just a few micrometers long. Flagella are long — up to 2 mm in length. The flagella of some insect sperm are this long. They are exactly the same from a biochemical perspective but were named differently because of how they look under the microscope.

Both of these structures have a base at the part where they stick out from the cell membrane. This is where the microtubule-based structure, called the *axoneme*, begins. The axoneme is the 9 + 2 structure with the 20 total microtubules in it arranged in 9 doublets plus 2 in the middle. This is a detailed view of the cross-section of the axoneme:

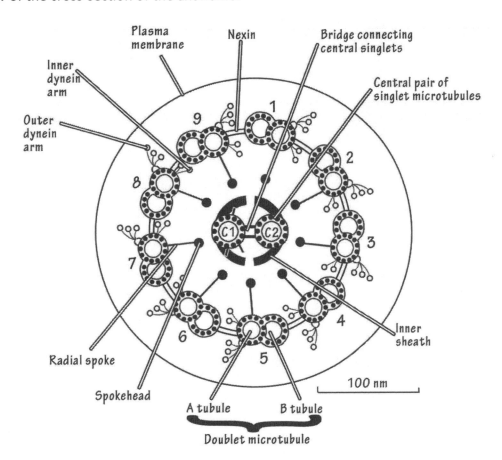

The basal body at the base of the cilia or flagella are basically modified *centrioles*, from which microtubules grow elsewhere in the cell. There are nine triplets of microtubules just like you'll see with centrioles. Above this structure is a transition zone that helps attach this to the rest of the axoneme. This is what the whole thing looks like together:

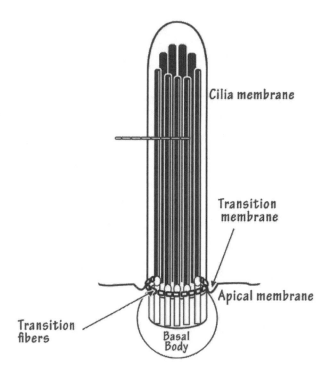

So, how do the cilia and flagella beat? First of all, they need energy. The energy, of course, comes from ATP. The rest comes from sliding filaments along one another over the entire axoneme in ways that can allow movement in any direction of these structures.

Like all parts of the cytoskeleton, the microtubules have no ability to move by themselves. They need to have some kind of motor protein to help do this. In the case of the flagella and cilia, it is largely *dynein* that does this. The dynein arms most likely reach out to the microtubules. The dynein arms bind to the microtubules, move the tubules along each other, and then separate in order to bind at a site further down. This is probably what it looks like:

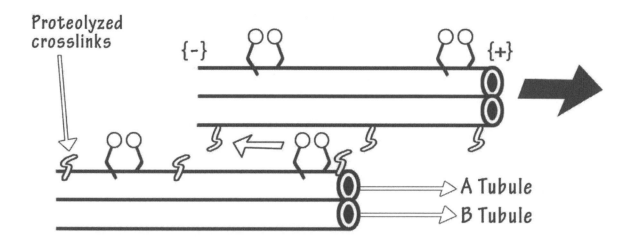

To Sum Things Up

There is real movement going on when the cytoskeleton interacts with motor proteins to allow for movement. A lot of internal movement happens when *actin* and *myosin* form bundles that slide across each other in ways similar to what happens in muscle cells. Things like *vesicular movement* and *cytoplasmic streaming* use either microtubules or actin/myosin bundles to allow organelles and vesicles to move, and to allow the circulatory flow of the cytoplasm to happen in certain cells. Amoeba and other motile cells use some version of *pseudopodia*, or "false feet," in order to allow the cell to move from one place to another.

Microtubules are the main fiber of the cytoskeleton to allow flagella and cilia to beat. These have a particular 9 + 2 arrangement of microtubules; these rely on the dynein motor protein to drive the ability of the microtubules to walk along each other, facilitating movement of the flagella or cilia.

CHAPTER 27:
CELLULAR DIGESTION

If you didn't catch it earlier, you should know now that *cell digestion* is not the same thing as *cell respiration*. In cell respiration, molecules are broken down in order to make energy. In cell digestion, however, there is breakdown of molecules, but the purpose is simply to recycle the parts. No energy production comes out of cell digestion.

The two major organs that participate in cell digestion are the *lysosomes* and *peroxisomes.* Lysosomes are the true garbage disposals for the cell. They take all types of molecules and break them up into monomers that can be ejected from the cell or recycled. These look like dense spheres in the cell of varying shapes.

There are about 50 different enzymes in lysosomes — all of them degrade something. Lipids, carbohydrates, RNA, DNA, and proteins can all be broken down with these enzymes. There are at least 30 different human diseases called lysosomal storage diseases caused by proteins that are defective, meaning they can't break down one or more molecule types.

Fun Factoid: The most common lysosomal storage disease is called Gaucher's disease. These patients have a mutated protein that doesn't break down glycolipids. These patients have enlarged spleens, enlarged livers, anemia, bruising, arthritis, and a risk for bone fractures. Another disease called I-cell disease means that all lysosomal enzymes can't get into the lysosomes to do their job.

These enzymes work best in very acidic environments, so the inside of the lysosome has a pH of about 5.0. It means that lysosomes have pumps that pump hydrogen ions into the interior of the lysosome so they can be more acidic as a whole. This is how it works:

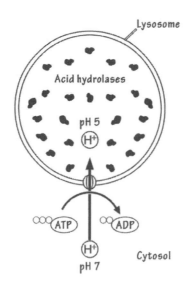

Lysosomes will take up digested material by the process of *endocytosis.* Lysosomes are initially made by the fusion of different transport vesicles that have budded off the Golgi apparatus. These bind to the endosomes coming from endocytosis. These fused lysosomes will start the process of molecular breakdown of anything trapped within these structures.

The process is complex, involving the uptake of substances from outside the cells into endosomes. These endosomes mature and then fuse with lysosomes before going to the Golgi apparatus to be processed. Similarly, the Golgi apparatus makes vesicles containing digestive enzymes used by these same lysosomes to do their job.

Phagocytosis, or "cell-eating," doesn't happen in all cells. The process happens in immune cells that recognize a pathogen, engulf it, and chew it up. Large pathogens like bacteria, cell debris, and old/dead cells can be engulfed and chewed up in structures called *phagolysosomes* (combined phagosomes and lysosomes). These can be very big and have the ability to chew up pathogens, spitting back out some of their proteins onto their cell surface to trigger responses in other immune cells. This is how it works:

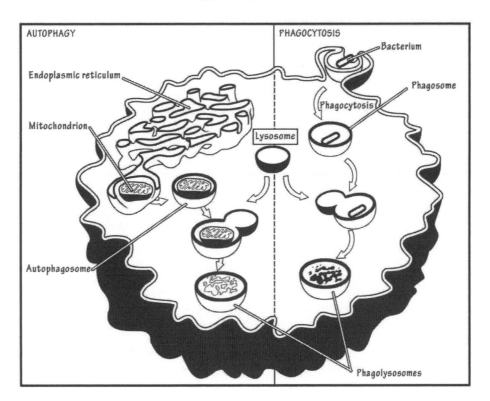

Autophagy is related to the process of phagocytosis. It involves chewing up organelles in the cell in order to get rid of them and to recycle their contents. In this process, an *autophagosome* is made that combines lysosomes with internal organelles. Within the autophagosome, the organelles get destroyed.

Besides enzymes, cells that participate in phagocytosis also have toxic molecules inside their phagolysosomes that help to destroy pathogens in non-enzymatic ways. Nitric oxide, hydrogen peroxide, and oxygen-based superoxide ions are all strong oxidizers that will directly detoxify and kill bacteria. There are NADPH-dependent oxidases that make these strong molecules specifically for the destruction of bacterial and dead cells, or cell parts. Children born with defects in NADPH oxidase have a high risk of bacterial and fungal infections because they can't destroy them through phagocytosis.

Peroxisomes

If we have such good lysosomes, why do we need to have *peroxisomes*? The peroxisomes and lysosomes look the same but are not made the same. Peroxisomes are actually assembled from proteins made in the ribosomes of the cell. The proteins are made as well and get imported into the peroxisome organelles. They are similar to mitochondria and chloroplasts because they divide during cell reproduction but don't have their own DNA.

Peroxisomes have a minimum of 50 enzymes that participate in biochemical pathways in these structures. They are good oxidizers, making hydrogen peroxide and similar molecules as part of their biochemistry. They have another enzyme called *catalase* that breaks down hydrogen peroxide so it doesn't later damage the cell.

Fatty acids, amino acids, and uric acid are just some of the molecules that can be broken down by peroxisomes. *Fatty acid oxidation* is especially important because it allows these molecules to be used later in cell respiration in order to make energy. In such cases, hydrogen peroxide is the breakdown product of fatty acid oxidation.

Peroxisomes can make lipids as well. In animal cells, both *dolichol* and *cholesterol* can be made by the peroxisomes in the same way that it happens in the smooth endoplasmic reticulum. They also make *plasmalogens*, which are phospholipids that can be used for the cell membrane of some cell types.

In plants, peroxisomes are more important in other ways. There are peroxisomes that take the fatty acids in the seeds, turning them into carbohydrates to help with germinating plants. Plants have a special cycle called the *glyoxylate cycle*; it is similar to the citric acid cycle and is the way peroxisomes can make sugars out of fatty acids.

In plants, peroxisomes help in photorespiration. They help to metabolize an end product of photosynthesis. Carbon dioxide goes on to make carbohydrates as part of photosynthesis in the Calvin cycle. Sometimes oxygen is accidentally put into the molecules in place of carbon dioxide. This makes the wrong molecules, including one called *phosphoglycolate*. This is a useless molecule, so it gets metabolized in peroxisomes to make the amino acid called *glycine*. Glycine goes on to make *serine*, which turns into *glycerate* in the chloroplasts so it can be put into the chloroplasts and into the Calvin cycle.

To Sum Things Up

Lysosomes and *peroxisomes* are two different vesicles that are involved in *cell digestion.* These are structures with the main job of destroying or breaking down larger molecules. Lysosomes have dozens of enzymes that break down all types of molecules. Some are sent to be recycled and others are packaged in order to leave the cell.

Phagocytosis is when pathogens and dead cells are engulfed by certain cells, usually immune cells. These will enter the cells as *phagosomes* that combine with lysosomes to make *phagolysosomes.* These together have the ability to break down very large things. Some of these pathogens have the parts of their membranes with proteins on them spit back out onto the cell membrane of the phagocytic cell, where the proteins become antigens for the immune system to do things like make antibodies.

The peroxisomes mainly oxidize molecules using a great number of oxidizing enzymes. The end product of this process is *hydrogen peroxide*, which is destroyed by another enzyme called *catalase.* Some of the products of peroxisomes engage in energy production. *Lipids* can be made or broken down in peroxisomes. When fatty acids are oxidized, they can be used in metabolism to make energy. In plants, they can recycle byproducts of abnormal Calvin cycle function in order to send them through the Calvin cycle again.

DIVISION FOUR:
A CLOSER LOOK AT GENETICS

This is the division where we talk about genetics from many perspectives as it relates to the cell. Nucleic acids may be the "blueprint" of the cell, but they can also be messengers of that blueprint, going on to make functional proteins. We will cover these first before talking about some even more interesting things, like how genes are regulated in the cell and how genetics is used for some amazing medical and agricultural technology.

SECTION EIGHT:
DNA, RNA, AND THE JOBS THEY DO

You now know what a nucleic acid is and what it looks like. In these next few chapters, we will talk more extensively about how they can be called *genetic material* or how they become *messenger molecules* for the making of the proteins all cells need. These are amazing molecules that are designed to provide a kind of alphabet that all cells can read and interpret in the making of proteins.

CHAPTER 28:
WHAT IS GENETIC MATERIAL?

Genetic material basically means nucleic acids. As you'll remember, nucleic acids are long polymers of nucleotides. There are eight different nucleotides in nucleic acids. Four of these are seen in DNA and four are seen in RNA. Just so you remember the details, let's look at the main differences between the two:

- **DNA** — These form very long polymers and are the main genetic blueprint in all cells (but not necessarily in viruses). These are double-stranded helices made from nucleotides, which are made of different nitrogenous bases, a sugar called deoxyribose, and a phosphate group. The different bases are adenine, thymine, cytosine, and guanine. This is what it looks like:

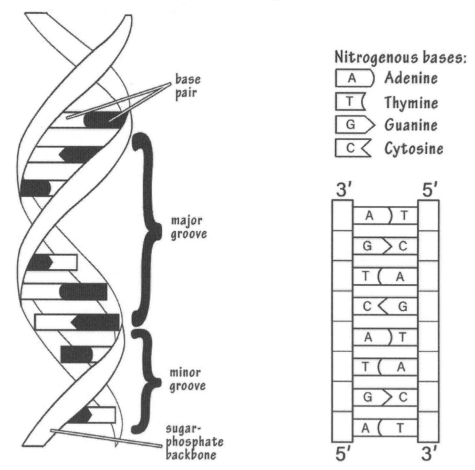

- **RNA** — These form shorter polymers than DNA. They tend to be single-stranded but will loop back on themselves to form temporary double-stranded segments. They come in several types that perform different functions. The RNA nucleotide is a nitrogenous base, a ribose sugar, and a phosphate. The different bases are adenine, uracil, cytosine, and guanine. Here's what it looks like:

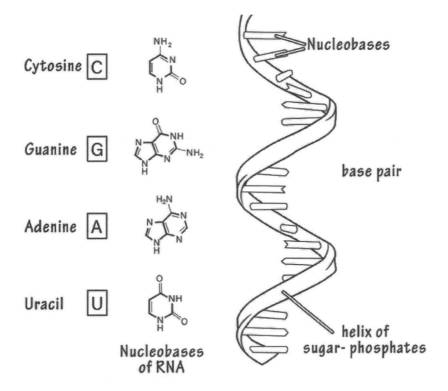

The most interesting feature of these nucleic acids is that they can create an entire alphabet with just four "letters," or bases. It's the arrangement of these bases that creates the "words" necessary to make proteins, of which there are 20.

If you consider the four bases in DNA and want to create an entire alphabet for these 20 amino acids, how many in a row would it take to have enough unique words out of these? It turns out that it takes three in a row. The total number of unique combinations you can make out of two in a row is 16 unique combinations. This isn't enough. Add a third one and you get 64 unique combinations. This is perfect because it's enough for every amino acid plus messages to say "start" and "stop" when transcribing proteins (and a few extra). Each set of three is called a *codon*.

The DNA and RNA are made from nucleotides. These are made themselves from a sugar molecule. In DNA, the sugar is a 5-carbon sugar called *deoxyribose* (because it just has a hydrogen atom at the 2' carbon atom). Ribose on RNA is the same but has a hydroxyl group there instead. This what these two look like:

The phosphate groups are the same in both molecules, but the bases are different. In DNA, there are only adenine and thymine that attach to each other through hydrogen bonding along the rungs of the DNA ladder. If RNA is made from a strand of DNA, the base that attaches through hydrogen bonding to adenine is *uracil*.

The bases that are purines have two rings on them. These are *adenine* and *guanine.* The *pyrimidines* have one ring on it; these are *thymine, uracil,* and *cytosine.* In DNA and RNA, if there is a hydrogen bonding attachment between the two, it will be between a purine and a pyrimidine. The GC combination of cytosine and guanine are stronger together because they have three hydrogen bonds between them, while the others have just two of these bonds. The base always attaches to the sugar at the 1' carbon on the sugar molecule.

Is it a nucleotide or a nucleoside? A nucleoside doesn't have a phosphate sugar, but a nucleotide does. This gets confusing because you can have things like ATP and GTP, which are nucleosides with three phosphate groups on them. You can also have ADP and GTP that have two phosphate groups on them. All molecules of this type produce the energy needed make nucleic acids.

When a nucleic acid polymer is made, the 3' carbon of the sugar molecule of one of the nucleotides binds to the phosphate group of another using what's called a *phosphodiester bond*. This is what it looks like when this happens:

These molecules have polarity but not in the same way that polarity is used in smaller molecules like water. What it means here is that the two ends are not the same. There is what's called the *5' end* and the *3' end*. In DNA, you should know that when the two strands come together, they do so in opposite directions. These are called *antiparallel strands*. The 5' end of one always matches with the 3' end of the other. This also affects the way DNA is made (because it is always made from the 5'end to the 3' end). When writing a DNA sequence, you'll always see them written from the 5' to 3' end. This is what it looks like:

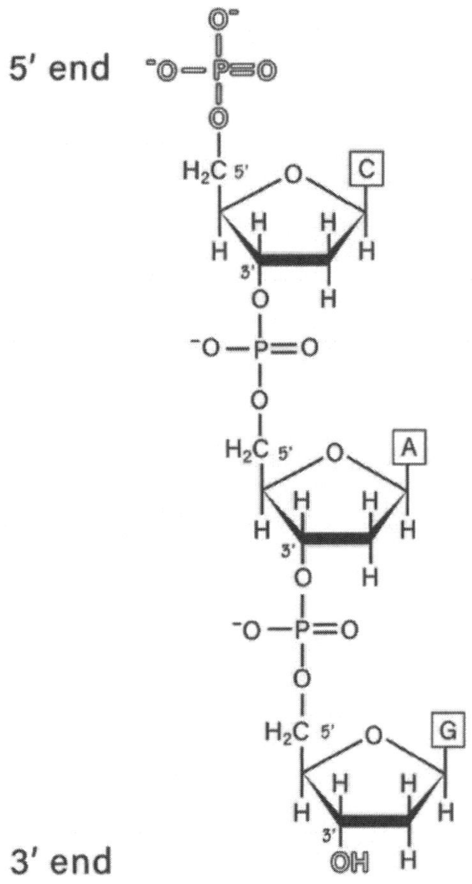

5' end

O⁻
⁻O—P=O
O

H₂C 5'   O
   H   H
 H   H
   3'
   O   H

⁻O—P=O
O

H₂C 5'   O
   H   H
 H   H
   3'
   O   H

⁻O—P=O
O

H₂C 5'   O
   H   H
 H   H
   3'
   OH   H

3' end

[C]
[A]
[G]

DNA strands must separate in order to be replicated, or "copied." This is why the bonds between the two strands are fairly weak. When they separate, the strands are now available to make new matching strands off of the split strands. They don't separate all at once but in sections. In these sections, the DNA strands are added to existing strands, forming two separate double-stranded DNA pieces. This is called *semi-conservative DNA replication* because one strand is old and one strand is brand new.

When the DNA gets unwound, there is a risk that the DNA molecule will be unstable enough to supercoil or twist in ways that make it difficult to put the DNA back together. This supercoiling also stresses the molecule. This is fixed by an enzyme called *topoisomerase*, which helps to stabilize the molecule so it won't break or stretch unnaturally. There is more than one type of topoisomerase, but this image shows basically what they do:

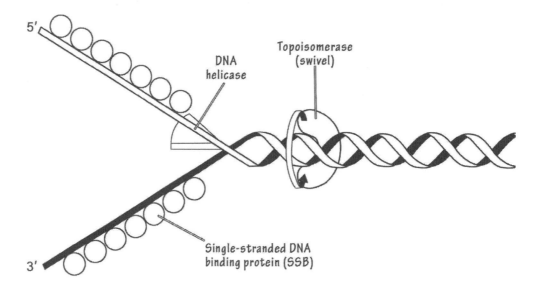

See how it binds to the DNA strands and encases them as they separate so they don't get too coiled up? As you'll learn later, the process of DNA replication involves more enzymes than this.

RNA Comes in Different Types

RNA is similar in many ways but there are many more types of RNA than there are DNA molecules. RNA can be single-stranded or double-stranded. It can also form a heterogeneous strand pair involving one strand of DNA and one strand of RNA. RNA strands are shorter and undergo different types of bending back in order to form what are called *hairpin turns*. The hairpin turns are exactly what they look like and are why certain types of RNA have a unique shape. Here is what RNA shapes can look like:

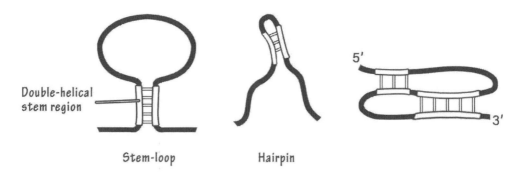

Stem-loop   Hairpin

There are different types of RNA molecules that do different jobs. This is a summary of the common RNA types and their basic structure or function:

- **Messenger RNA** — This is the one that takes the DNA message and sends it to the ribosomes to be made into proteins. Surprisingly, only five percent of all cellular RNA is of this type. It has the ability to have the complementary genetic code on DNA so the message gets passed to the ribosomes. There are three bases per codon and 64 possible codons. There are 20 amino acids and some have different codons that make the same amino acid. There are several start codons and three stop codons.

- **Ribosomal RNA** — This is called *rRNA* and makes up 80 percent of all the cellular RNA. These are large pieces made in the nucleolus. They travel to the endoplasmic reticulum and combine into two halves (called the 50s half and the 30s half in eukaryotes). They need to combine with proteins in order to make ribosomes that work to make polypeptide chains.

- **Transfer RNA** — *Transfer RNA*, or *tRNA*, is small. Its main job is to grab a single amino acid in order to attach it to the growing polypeptide chain in the ribosomes. This is why they are called transfer RNA. There is a specific tRNA molecule for every one of our 20 amino acids. These are interesting molecules because they have a lot of hairpin turns that allow them to have a unique shape. This is what one of these molecules looks like:

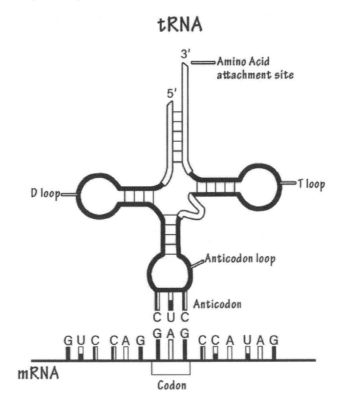

- **Small nuclear RNA** — This is also called *snRNA*. It is used in the nucleus in order to take a large section of messenger RNA and cleave out the segments not necessary for the final molecule. *Introns* are sections that get cut out, while *exons* remain and get made into the final messenger RNA molecule in order to allow it to leave the nucleus, headed for the ribosomes.

- **Regulatory RNA** — This is called *aRNA, siRNA,* or *miRNA,* which stand for *antisense RNA, small interfering RNA,* and *microRNA,* respectively. The miRNA can help to accelerate the breakdown of RNA so it can't get translated into proteins. The siRNA can be made from viruses that have RNA in them. They can also block an action or messenger RNA. They have segments called *riboswitches* that help regulate messenger RNA activity.

- **Transfer Messenger RNA** — This is also called *tmRNA*. It is found in plastids and bacteria. The job of this type of RNA is to tag certain proteins made by mRNA molecules. They help certain mRNA molecules that don't have the stop codons they need in order to allow the polypeptide chain to break off after a protein has been made.

- **Ribozymes** — These are RNA-based enzymes that act just like protein-based enzymes because they catalyze reactions.

To Sum Things Up

Nucleic acids are polymers of *ribose-based nucleotides* or *deoxyribose-based nucleotides*. They work to provide the genetic material for the cell or they copy themselves according to the DNA message so that proteins can be made. DNA is mostly double-stranded except where they are getting replicated. RNA segments are smaller. They can do many things, including taking the DNA message off of a gene segment, processing the mRNA in some way, and helping polypeptide chains get made by putting one amino acid on them at a time.

CHAPTER 29:
THE REPLICATION OF DNA

In this chapter, we will talk about how DNA is made, or replicated. One of the definitions of "life" is that it is can reproduce itself. DNA does this nicely through *semi-conservative replication.* As you remember, this means that each daughter cell of a cell that divides into two will have one strand of DNA coming from the original parent cell, while the other comes is made anew during the replication process.

It's best to see the process as a series of steps and look at how each step works by itself. There are many enzymes involved in making DNA that we will talk about as they come up. By the end of these steps, you should have a good grasp of how DNA is made when a cell divides.

Step 1: DNA is unzipped – There is an enzyme called *helicase* that breaks the hydrogen bonds between the base pairs so that the double-stranded DNA becomes single-stranded DNA. This happens all along the DNA molecule, leaving a Y-shaped area called the *replication fork*, which is where the template strands (the original strands) get added by putting new base pairs along the strand one at a time. This is what helicase looks like when it acts on the DNA molecule:

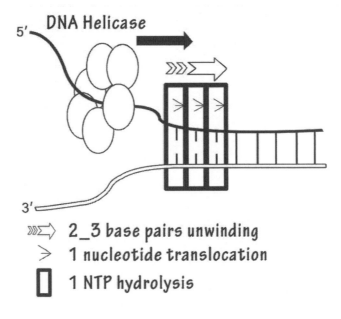

There are other necessary enzymes for this process. We already talked about *topoisomerase,* which helps stabilize the DNA strands so they don't get supercoiled or stressed while they are single-stranded.

Step 2: A leading strand and lagging strand are made – The leading strand goes in the 3' to 5' direction, or toward the replication fork. The lagging strand is oriented in the opposite direction, away from the replication fork. This is what it looks like:

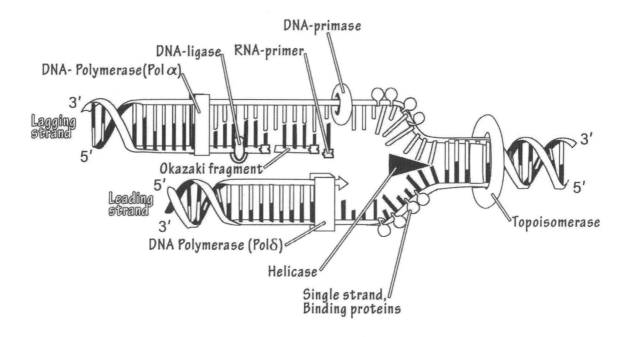

The leading strand is called "leading" because the strand can be made continuously so that the new strand is built from the 5' to 3' direction, which is the only way DNA can be made. The laggings strand is called "lagging" because DNA cannot be made backward. Instead, small fragments called *Okazaki fragments* are made in the proper direction but in pieces along the lagging strand.

The enzymes involved in this include *DNA primase*, which are also called *RNA polymerase III*. Its job is to make a short section of complementary, or "matching," RNA in order to allow DNA polymerase to work. DNA polymerase is what makes the DNA section you are looking for, but, because it cannot just start making DNA from scratch, it needs a primer strand of RNA to give it a boost. This is what primase does.

Then, DNA polymerase acts to make the new DNA segment. This is an amazing enzyme because it has its own proofreading function. It proofreads itself as it goes along so that the error rate is very low. It needs to be low because you wouldn't want many *mutations*, or "errors," made that would get passed to the next generation of cells. The error rate is between one in 100 million and one in a billion mistakes for every nucleotide laid down.

The lagging strand DNA is still laid down in a 5' to 3' direction but it will make multiple discontinuous Okazaki fragments. These need to be joined up if the DNA on that strand is to be continuous in the end. This is done using an enzyme called *DNA ligase*. The entire purpose of DNA ligase is to connect these broken Okazaki fragments on the lagging strand.

The end result is two separate but identical chromosomes of double-stranded DNA. These are then separated into the daughter cells in the process of *cell division*. We will talk about how cell division happens later on so you can see how all of these processes come together to make two identical daughter cells through what's called *mitosis*.

To Sum Things Up

DNA replication is how DNA copies itself prior to having the entire cell divide into two daughter cells. The process is semi-conservative so that both DNA strands act as a template for a complementary strand that is made before a cell divides.

There are many enzymes necessary for DNA replication to occur. These include *helicase, topoisomerase, primase, DNA polymerase,* and *ligase.* These work together to allow the strands to separate and get copied on the leading and lagging strands. There is an extra step to the lagging strand section, which is to connect the short Okazaki fragments that get made on the lagging strand but not on the leading strand.

SECTION NINE:
HOW CELLS DIVIDE

This section will talk about how different types of cells divide. The three types of cells we will talk about include prokaryotes, eukaryotic somatic cells, and eukaryotic germ cells. Prokaryotes like bacteria have their own special way of dividing called binary fission.

Eukaryotic somatic cells (in animals at least) are those that just divide in order to increase their number. These include your liver, kidney, and epithelial cells; they divide in order to replace dead cells or to grow the organ or tissue. *Germ cells* are the cells that make the sperm and egg cells of animals (although plants do this too).

We will also talk about cell communities. Prokaryotes have their own cell communities, which are sometimes called *biofilms*. These cells communicate with one another in order to allow the whole community to have a better chance of survival. Cell communities in multicellular eukaryotes like humans are also called *tissues*. We will talk about the types of tissues and how the cells in them communicate with one another.

CHAPTER 30:
WHAT IS CELL REPRODUCTION?

If we decide to define cell reproduction in the most basic way possible, we will say that cell reproduction is when a cell divides to make two copies of itself. In such cases, the initial cell is called the *mother cell* and the two cells it makes are called *daughter cells.*

In reality, cell reproduction is more complicated than just one cell making two daughter cells. In *meiosis*, for example, one cell can make four separate and unique *gamete cells*. A gamete cell is either a sperm or egg cell. They only have half the total number of chromosomes as the adult, or *diploid* cell, so they can combine with other gametes to make a whole new *diploid zygote.* The zygote is a single cell that will be able to divide and differentiate into a whole organism.

Organisms must divide in order to be able to survive as a species. They need to divide in order to have offspring pass on their genetic legacy. With *mitosis* and *prokaryotic binary fission,* the idea is to have the genes in the offspring be identical to the parent cells. This is effective, but it doesn't allow for any genetic diversity. If the environment changes somehow and the offspring are not diverse enough that some survive, the whole species will die off.

This is the potential benefit of sexual reproduction, or the *meiotic process.* In meiosis, there is an interesting feature called *genetic recombination* that happens. It means that the genes are rearranged in the process to make unique gamete cells. This is a good idea in situations where diversity is important. By having a diverse population of offspring, the species as a whole can survive in situations where the environment might change; in that kind of event, there is a good chance that some of the offspring will be able to adapt to the changing environment.

When a cell divides, its genetic material is copied. In cells, this means its DNA is copied. The cell grows before dividing, so both daughter cells have enough *cytoplasm* and important organelles like *mitochondria* or *chloroplasts* to divide for the same reason. Things like the endoplasmic reticulum often get broken down before cell division but get remade after cell division has taken place.

Types of Cell Reproduction

The two major types of cell reproduction are *asexual reproduction* and *sexual reproduction.* With asexual reproduction, there is just one cell or organism that creates a copy of itself. The daughter or daughters have the same genetic material as the parent cell. There is just one parent.

In sexual reproduction, there are two parents involved. These two parents offer up some of their DNA in order to make a combined offspring organism that has characteristics and the genetics of both parents combined.

This is basically what these two types of reproduction look like:

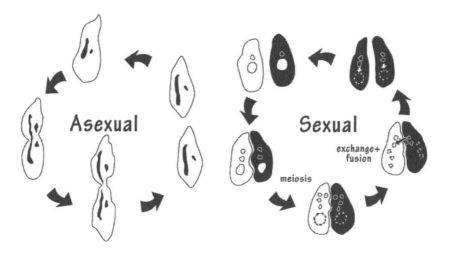

Prokaryotes mostly divide through *binary fission* (more on this later). This is not the same thing as mitosis. The only similarity between the two is that the offspring cells are genetically the same as the parent cell. Remember that prokaryotes have *circular DNA*. This circle gets copied and divided so that the offspring/daughter cells have a whole new piece of their own DNA.

Budding is also an example of asexual reproduction. This can happen in certain fungi, hydra, and yeast organisms. These cells will break off a tiny piece of themselves. Before the bud comes off completely, the nucleus is copied entirely and one migrates to the budding site. The bud then gets pinched off and acts like a spore that can later grow and develop into a bigger (but identical) offspring organism.

This is what yeast budding might look like:

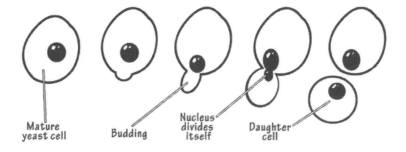

Higher organisms (*eukaryotes*) will divide through *mitosis*. This is more complex than binary fission but still requires making a faithful copy of the parent cell before dividing into two daughter cells. The DNA is divided into chromosomes instead of being circular, so the chromosomes must have a fancier way of both duplicating themselves and then dividing into daughter cells. These daughter cells are identical to the parent.

Finally, there is sexual reproduction, which always involves meiosis in eukaryotes. This is different when the organism is male versus a female, even though the process is identical in most respects. In humans, for example, one *spermatogonium* divides in two stages to make four non-identical sperm cells. In females, the *oogonium* is the germ cell; it divides in an identical way but makes only one egg cell. You'll see how this happens later. In both cases, the gamete is a *haploid cell* with half of the total chromosomes the adult cell needs.

Binary Fission

Binary fission is what prokaryotes like bacteria reproduce. It is actually fairly simple. First the cell must get bigger — twice its original size. Then, it must copy its single circular chromosome before dividing into two. All of this takes a little bit of finesse because it must divide when there are enough nutrients and when the environment is right. If the cell is in a community, it must communicate with other cells so it divides (or doesn't divide) based on the whole community's needs and not the just the needs of the single cell.

This is what binary fission looks like:

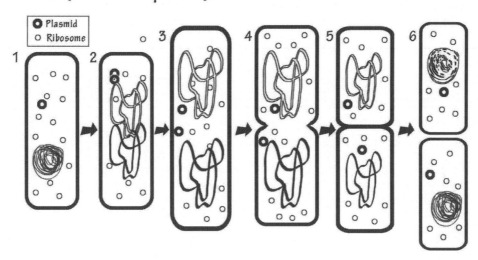

In the first step of binary fission, the cell needs to make a copy of its genetic material. The single large chromosome must copy itself. If the cell has any *plasmids* (secondary tiny circles of DNA), these might or might not copy themselves. It is not necessary that they do this, though. If one plasmid doesn't copy itself, it will only show up in one daughter cell.

There are proteins that must be present for cell division to happen. One protein, called FtsZ in bacterial cells, comes together in the middle of the cell so that the cell divides cleanly without damaging the two separate circles of DNA. Once this happens, these cells have walls that also need to be made. All of these steps (*chromosome replication, chromosome division/separation, cytoplasm division*, and finally *cell wall formation*) happen in the same order and in a highly regulated way.

Bacterial Budding

Bacteria will bud, as well as yeast and fungi. Not much is known about bacterial budding except that it happens in much the same way as it does in yeast cells. There are bacteria called *segmented filamentous bacteria* (not very common types) that reproduce only in this way. This is how a bud forms off of a bacterium:

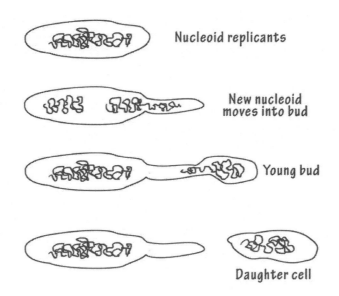

To Sum Things Up

Without the ability to reproduce, a species of organisms (both small and large) will not survive. Reproduction happens in all forms of life in *sexual* and *asexual* ways. Most of the asexual reproducers are bacteria and other prokaryotes. They almost always divide through *binary fission*, although a few types will divide through budding off pieces of themselves.

Budding happens more regularly in yeast cells and other fungal organisms. The cell breaks off a piece of itself, called a *bud*, and places a duplicated nucleus in the bud before it separates. This bud can sometimes stay the same small size until the environment is right. When this happens, the bud will grow into a new cell.

All eukaryotic cells participate in *mitosis*, which is also *asexual reproduction*. The cell will duplicate its DNA in chromosomal form after it grows, so that the daughter cells can have enough cytoplasm per cell. Mitosis is a series of events that cause the cell's chromosomes to separate before the entire cell divides its cytoplasm and other organelles. The end result is two identical daughter cells.

Meiosis is where the cells participate in sexual reproduction. Sexual reproduction involves two separate parents making *haploid cells* so these cells can later come together to make a unique *diploid offspring cell.* The advantage to this type of reproduction is that the offspring have a better chance of helping the species survive if the environment changes somehow.

In this next chapter, we'll talk mainly about the cell cycle of eukaryotic cells and how they ultimately divide to make identical daughter cells through the process of mitosis.

CHAPTER 31:
THE CELL CYCLE AND MITOSIS

The cell cycle is what each cell must go through in order to divide into two daughter cells that are identical to the mother cell. As this cycle progresses, the cell grows, duplicates its chromosomes, and then separates them into two daughter cells.

You've probably heard a lot about mitosis and might think that this is all cells actually do. This is far from the truth unless the cell is a cancer cell. Cancer cells that are particularly aggressive will divide over and over again without really taking much of a break. Most cells are in mitosis less than 10 percent of their total lifespan.

Cells in the human body vary greatly as to how much they spend in mitosis and how much they spend outside of it. There are cells that essentially never divide — like nerve cells in the brain. Other cells, like the epithelial cells lining the GI tract, will divide more frequently to make up for cells that get lost in everyday processes.

This is what you'd see in terms of a cell cycle in the average cell:

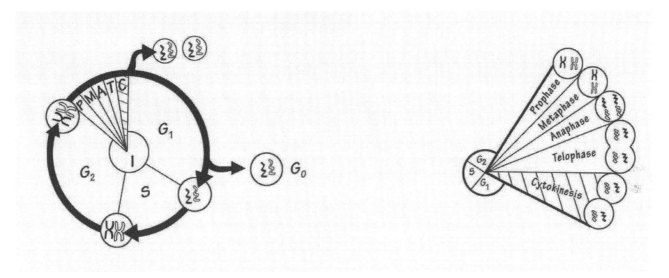

The two basic sections of the cell cycle are called *interphase* and the *mitotic phase*. Don't think that essentially nothing happens in interphase just because you can't see it happening. In fact, besides operating the usual cell machinery during interphase, the cell also does its DNA copying, or *replication*, in this phase. The DNA only separates during mitosis — it doesn't divide because it has already done this in interphase.

You can imagine that, since some cells don't divide at all and others divide frequently, there must be some control over how and when it happens. In fact, there are controls, which are called *cell cycle checkpoints.* These happen throughout the cell cycle in order to halt the process or allow it to happen unchecked, depending on the circumstances.

Phases of the Cell Cycle

There are four or five cell cycle phases, depending on how you look at it. If a cell is dividing in any way, there are just four phases: the *G1 phase, the S phase, the G2 phase,* and the *M phase.* We'll talk about each of these phases in detail. Cells that never divide and are totally quiescent are in a phase called the *G0 phase,* which is essentially never-ending in some cell types. You can see in the above image where the G0 cycle is located.

Notice that a cell can technically jump into and out of being quiescent versus actively dividing. This only happens between the G0 phase and the G1 phase. Once a cell is in the S phase or beyond, the possibility of going back and becoming quiescent is lost. The cell can certainly stop progressing into cell division after the S phase but it won't enter the G0 phase during those times.

Let's look at these phases separately:

- **Gap 0 or resting phase** — This is what we call a cell that is not actively dividing at all. The cell in this phase isn't in the cell cycle at all because it isn't cycling. No new cells are made, but the possibility of going back into the G1 phase is always there in most cell types. Cell types in humans vary greatly in whether or not they are always, mostly, or rarely in this G0 phase. There are cells called *post-mitotic cells* that are either totally quiescent, or *senescent,* meaning "old." A post-mitotic cell is one that isn't dividing for whatever reason. Some cells can be too stressed or damaged to divide, so these are not dividing even though they aren't in the G0 phase.
- **Gap 1 or first interphase step** — This is when the cell increases in size. You won't see anything different by looking at the cell, but at the end of this phase there is a checkpoint called the *G1 checkpoint.* A cell needs to be cleared at this point before DNA can be copied in the S phase. The cells in this phase are very active biochemically, making new proteins and doing what they need to as part of whatever cell type. This is the time for increasing the numbers of organelles so that each daughter cell will have enough of these when the cell divides. Any cell in this phase can get out altogether to become a G0 (resting) cell. It can also simply stop and wait, or it can go on to the S phase.
- **S phase, or "synthesis"** — The cell that gets past the G1 checkpoint goes into the S phase. The S phase is where the DNA in the chromosomes undergoes replication. After the phase is over with, the cell will have the same number of chromosomes, but each chromosome will have a pair of DNA strands called *sister chromatids.* These are identical strands of DNA connected to one another. This is what two chromosomes look like during and after this phase:

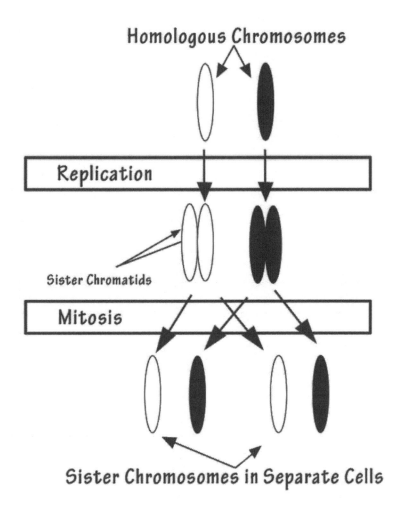

Homologous Chromosomes

Replication

Sister Chromatids

Mitosis

Sister Chromosomes in Separate Cells

The S phase is so busy doing replication that it doesn't undergo much protein-making or other activities. The only proteins made to a great degree are the *histone proteins*, which are needed to wrap pieces of DNA up to keep the strands from being so long.

- **Gap 2 phase** — This is also called the *G2 phase.* It starts as soon as the S phase is over. The cell keeps growing and prepares itself for mitosis and cell division. It won't pass this stage until a second G2 checkpoint is passed. The cell grows quickly in this phase and microtubules get made. Remember that microtubules are needed for mitosis to happen. There is a regulatory protein called *p53* that is a big part of the G2 checkpoint. The p53 protein will repair any damaged DNA in preparation for cell division. It will also cause the cell to die if DNA repair is not possible.

Fun Factoid: This p53 protein is super-important. Most cancers in the human body happen in cells where the p53 gene is damaged. The p53 protein is called a "tumor suppressor protein" because it prevents tumors. A disease called Li-Fraumeni Syndrome is hereditary and means a person has a high cancer risk because none of their cells have a good p53 protein.

- **M phase, or Mitotic phase** — No cell growth happens in this phase as the cell is busy dividing. The cell has a third checkpoint as part of this phase, so even when the cell seems totally committed to dividing, the whole thing can come to a screeching halt even this late in the game.

Mitosis

The *mitotic phase*, or *M phase*, is interesting to look at under the microscope. This is the only time where the chromosomes are completely visible. This is because they condense during this time, making them thick or fat enough to see under the microscope. There are two broad areas of cell division by this time: 1) *karyokinesis* and 2) *cytokinesis*. Mitosis itself is *karyokinesis* because it means "dividing the nucleus." After mitosis, there is *cytokinesis*, which means "dividing the cell." These are two separate processes that finish the cell division steps.

The different steps of mitosis are labeled according to what the cell looks like and what happens in the cell during each step. Let's take a look at these.

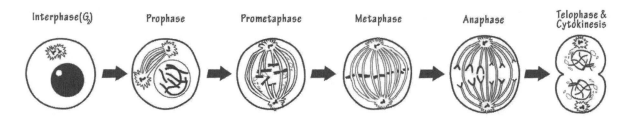

(If you need to memorize these steps in order, just remember "**p**ee on the **mat**," which is a way of remembering *PMAT*, or prophase, metaphase, anaphase, and telophase).

In *prophase*, there is the beginning of mitotic spindle formation involving *centrosomes* and *microtubules*. There aren't any chromosomes attached to it yet, though. These are busy condensing, while the nucleolus in the nucleus disappears. Late prophase is also called *prometaphase.* This is when the nuclear envelope around the chromosomes breaks down, too. You can easily see the chromosomes because they are very condensed. The microtubules in the mitotic spindle start to capture the chromosomes in order to prepare for the next phase.

Chromosomes look interesting in this phase. They have sister chromatids attached roughly in the middle by a *kinetochore*, which is what helps the microtubules separate when the time comes. It looks like this:

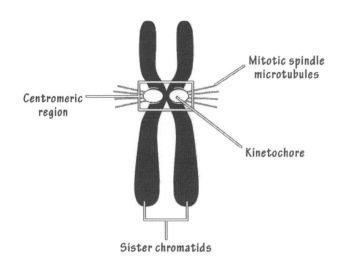

You might be confused about *centromeres* and *kinetochores*. The centromere is the part of the chromosome that holds all the sister chromatids together after the chromosome has been copied. The kinetochore is in the same region but is a disc of proteins that helps the microtubules of the mitotic spindle attach to the sister chromatids.

The centromere is a piece of the chromosome itself that defines where the sister chromatids are held together. It has no histone proteins but is where the kinetochores attach. Most organisms have just one centromere (except for some worms that have more than one). While the centromere is made of DNA, the kinetochore is made of proteins. It binds to the centromere to form a complex where the microtubules can attach. The kinetochore is layered, so parts attach to the chromatids and parts attach to the microtubules. There is a middle layer but no one knows its purpose. This is what the chromatids look like as they get ready to separate:

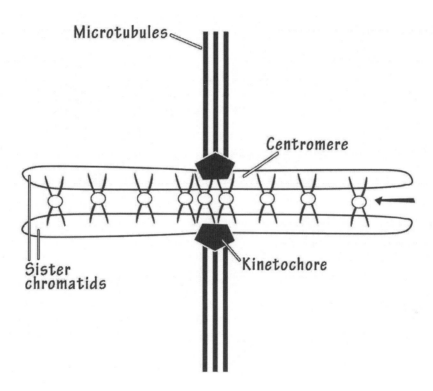

The centromere doesn't attach to the microtubules itself. It relies on the attached kinetochore proteins to do this kind of attachment so the chromatids can divide.

Not all microtubules attach to the kinetochores. Those that do are called *kinetochore microtubules.* The others attach to the cell membrane on the opposite sides of the cell. This is necessary to keep the whole thing stable and helps the chromatids separate appropriately. The microtubules that start at the centrosomes on either side of the cell and attach to the membrane are called the *aster* because they form a star shape.

The next step is called *metaphase*. This is where the chromosomes line up on what's called the *metaphase plate.* The metaphase plate isn't a real thing but is rather an imaginary line where the chromatids are held taut by the mitotic spindle to get ready for separation. The chromatids are completely captured and able to be manipulated by the spindle.

There is a checkpoint at this stage. What happens here is that the chromosomes are checked to make sure they are properly lined up and that they will separate cleanly. If this isn't possible, the cell won't continue mitosis past this point. They call this the *spindle checkpoint.*

The next step is *anaphase*. Think of this phase as being when the microtubule spindle revs its engine and starts to pull these chromatids apart. There are "pulling" microtubules that draw the chromatids to opposite poles of the cell. There are also "pushing" microtubules that push between the chromosomes to help the process move along.

Once the glue that holds the chromatids together breaks down and the DNA separates, the DNA strands are no longer called *chromatids*. They are just called *chromosomes* now and are busy getting pulled and pushed apart. Remember that this whole process involves motor proteins that drive the process of chromosome separation.

The two proteins involved are the *kinesins* and the *dyneins.* The kinesins help to position the chromosomes as they separate. The dynein proteins regulate the length of the spindle as the chromosomes are separating. Remember that microtubules are just tubules. They cannot pull anything apart by themselves.

The next step is *telophase.* The chromosomes are now separated and start to decondense, so you won't see them after this. The nucleolus starts to develop on the two cell sides and the nuclear membrane captures both sets of chromosomes into two nuclei. The spindle eventually disappears as well, leaving behind one centrosome for each daughter cell.

Cytokinesis is not really a part of mitosis but happens shortly thereafter. It's slightly different in animal and plant cells. It looks a bit like this:

CYTOKINESIS

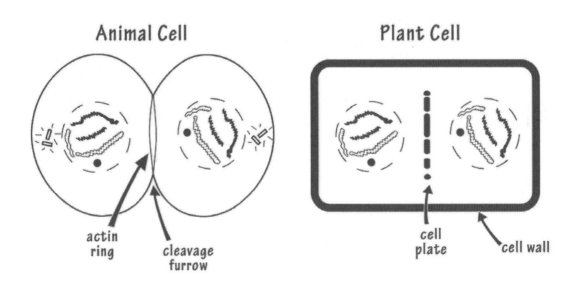

Animal Cell   Plant Cell

actin ring   cleavage furrow   cell plate   cell wall

In animal cells, there is an *actin* (microfilament) *ring* that cinches the cell like a purse string so that the two cells are split off from each other. In plant cells, there is a *cell plate* formed instead. The cell plate is made from cell wall material; it helps separate the two cells along with the cell membranes of the daughter cells that form at the same time.

Cell Cycle Checkpoints

If you've kept track so far, you know that there are three separate checkpoints in the cell cycle: *G1 checkpoint, G2 checkpoint*, and the *metaphase checkpoint*. There are special proteins called *cyclins* and *cyclin-dependent kinases* that help these checkpoints happen. The cyclins are the main regulatory proteins, and the kinases, or *CDKs*, are the enzymes that cyclins bind to in order to allow the cell cycle to continue.

This is the general idea for the G1 checkpoint: There is a signal from outside the cell, called a *pro-mitotic signal*. Then, the cyclin-CDK complex gets activated. Remember that kinase enzymes add phosphate groups. When this happens, transcription factors and DNA replication enzymes get activated and any inhibitor proteins get destroyed. The cell is then allowed to copy its DNA in the S phase.

Each checkpoint works in roughly the same way but allows for different things to happen during the cell cycle. There are different cyclins at work each time. If you measure the different cyclin concentrations at the various checkpoints, you get a graph that looks a lot like this:

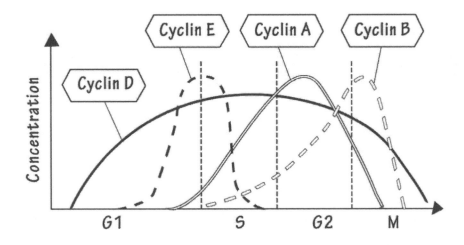

Fun Factoid: If you could make a good anti-cancer drug, what would it look like? Think about it. If you had a drug that blocked the checkpoints by mimicking the checkpoint blockers in the cell, you could slow down cancer growth by causing the cancer cells to also have their checkpoints blocked. That would be genius and is exactly what doctors and researchers are looking at!

The G1 checkpoint is the main one the cell uses to control its cell cycle. It is so important that it's called the *restriction point* of the cell cycle. The cell checks to make sure it has all it needs to continue dividing, such as the *DNA synthase enzyme* and *nucleotides*. If a cell is malnourished, it won't be able to progress.

The G2 checkpoint is where the cell looks to see if it has the cytoplasm and cell membrane components for the two daughter cells. There is also some kind of external timing mechanism that helps decide if it is the right time to divide or not.

The metaphase checkpoint is strictly to see if the metaphase plate situation is okay. If the spindle is properly formed and the chromosomes are lined up correctly, the checkpoint is passed and the cell completes the process of cell division.

Apoptosis Is "Programmed Cell Death"

It wouldn't be fair to talk about the cell cycle without also talking about how cells die. Certainly, a cell can be injured or infected and can then rupture, leading to its death. *Apoptosis* is not the same thing. The cell never ruptures but instead just withers away and dies. It happens all the time in multicellular organisms that need to have some cells die off as part of growth and development. Because it is often a planned thing, it is called *programmed cell death.*

Fun Factoid: *Apoptosis is both very common and necessary. Did you know, for example, that half of all nerve cells in the developing brain die by apoptosis almost immediately? As adults, many billions of cells in your body die off this way each hour of your life. Most of these cells die off even though they are completely healthy at the time of their apoptotic death.*

A baby's fingers are largely formed because of apoptosis. They start out looking a lot like fish flippers but then become formed into thin fingers as the unwanted parts die off. Tadpoles turn into frogs because of apoptosis of the tail. The nerve cells in the brain adjust their numbers to fit the number of target cells needing innervation. It's how these important developmental changes are allowed to occur. This is what it looks like this when it happens:

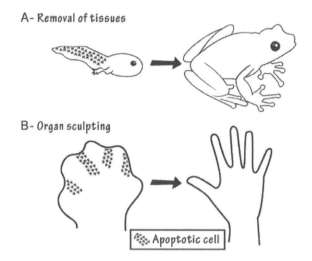

We adults would either grow or shrink if the number of cell divisions didn't get balanced by the number of cell deaths through apoptosis. How is this process so successful?

The cell undergoing apoptosis simply shrinks and becomes condensed. The entire cytoskeleton collapses on itself and the nuclear envelope disappears. The DNA breaks up into pieces and the cell puts out receptors indicating it needs to be eaten by your immune cells that recognize a dead cell.

Nothing leaks out of the cell and no inflammation occurs like you'd see with a cell dying after an injury. A cell that dies after an injury is instead said to have *cell necrosis.*

There are enzymes called *caspases*, which do the majority of apoptosis on a cellular level. These are *proteases* that target the *aspartic acid amino acids* in the polypeptide chain. There are pre-enzymes called *procaspases* that stay inactive until they are cleaved by other caspases. As you can imagine, this sets up a huge cascade of activated enzymes that cause all the proteins in the cell to break down. There are also enzymes that break down the nuclear envelope and DNA segments.

Essentially, the procaspases are there the whole time — like a series of time bombs just waiting for the right moment to be set off. There are adaptor proteins that drive the first series of procaspases together so that they are close enough to cleave each other's polypeptide chains. These then get activated, starting the avalanche that follows. The end result is a dead and shrunken cell.

Cells can be triggered from the outside by having death receptors outside of them. These receptors are attacked by your immune system, which has cells making ligands that ultimately start the cell apoptosis process. Through the cell signaling processes you already know about, the first procaspase molecules get activated and the cell is destroyed. Other cells just kill themselves by starting up the process internally.

This is how it happens if external factors are involved:

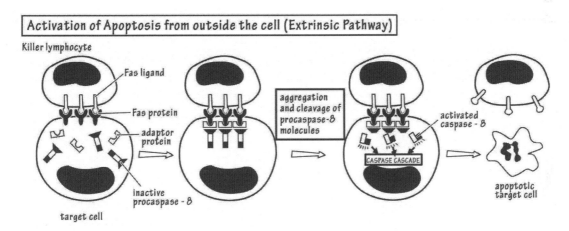

This is how it happens if internal factors are involved:

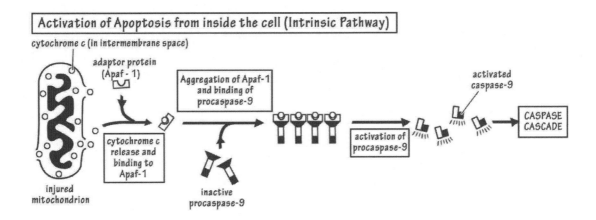

So, regardless of the trigger, the end result is the same. The cell shrinks down and the immune system notices the dead cell, eating it like any other dead thing it finds in your body.

To Sum Things Up

The *cell cycle* is a map of the cell's lifespan. A nondividing cell is said to be *quiescent*, or resting, but it can also be old, or *senescent.* Cells that are dividing go through four stages: G1, S, G2, and M (in that order). There are three checkpoints in the cell cycle that make the decision as to whether or not a cell will continue dividing.

Mitosis is a small part of the cell cycle but is interesting because you can see it under the microscope. The DNA of the cell has been copied by this time, so the job in mitosis is to simply separate the sister chromatids. The end result is that the nuclear material divides; the cell separates into two daughter cells in the process of *cytokinesis.*

CHAPTER 32:
MEIOSIS

Meiosis is the cellular basis for sexual reproduction. It is how *germ cells* (which are different from your average somatic cells) prepare the organism for producing the next generation by making *haploid gamete cells* to contribute to their offspring. There are equal contributions of *nuclear DNA* by both parents but the vast majority of *mitochondrial DNA* comes from the egg cell (and you'll see why in a minute). In Greek, meiosis means "diminution," referring to the shrinkage of chromosome numbers in this process.

Meiosis is much more complicated than *mitosis,* largely because it involves two separate sections. These are called *meiosis I* and *meiosis II.* Let's look at some terminology first.

The 22 pairs of similar, or *homologous*, chromosomes in humans are called *autosomes*. These are numbered 1 through 22. The reason these autosomes are homologous is because they are similar; they have the same basic genes — just two versions of each one. The other two are *non-homologous* because they have no similarities to one another. They are the *sex chromosomes*; in humans and most other organisms, they are labeled *X* and *Y*.

In meiosis, gametes are haploid. They do not have the full set of 46 human chromosomes (in humans); instead, they have 23 single sets. They don't get 23 random chromosomes. They get one of each kind, plus either an X or Y chromosome.

When a cell is just doing its thing, the two homologous chromosomes are not attached to each other and might not even be near one another. If a chromosome is copied, though, the copies are called *sister chromatids* and are connected to one another through the centromere. As we already talked about, they don't become chromosomes again until they separate in anaphase.

Meiosis I

For the purposes of meiosis, the homologous chromosomes must find each other in that mess inside the nucleus. How they do this isn't well understood. There are probably complementary base pairs on each of the homologs that help them find each other.

Before they can even pair up, they must copy themselves in the same way they do in mitosis. Once they do pair up, it is even stranger than in mitosis because now it is four chromatids connected to one another. This is known as a *bivalent*. The first prophase period is long in meiosis, mostly so that the homologs can find each other. It can take days or years for this process to happen. You'll see soon that this is where recombination occurs, which is another process that cannot be rushed.

This is basically what meiosis I looks like. It involves homolog pairing after duplication of the strands (with four sister chromatids), recombination, and all the stages of mitosis that happen after that — *metaphase, anaphase, telophase*, and cytokinesis. There are two daughter cells as a result; these are not identical but are still diploid cells.

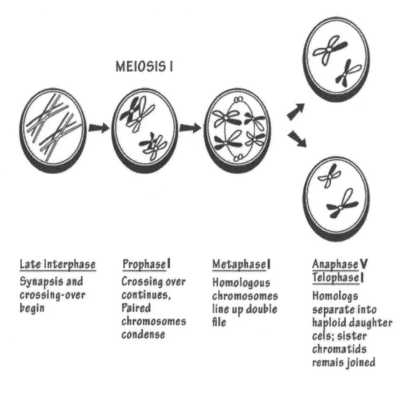

MEIOSIS I

Late Interphase
Synapsis and crossing-over begin

Prophase I
Crossing over continues, Paired chromosomes condense

Metaphase I
Homologous chromosomes line up double file

Anaphase V Telophase I
Homologs separate into haploid daughter cels; sister chromatids remais joined

Meiosis II

This is the part where the four haploid gametes get made. After meiosis I, there will be two diploid daughter cells. This is where it diverges even more from regular mitosis. The number of chromosomes is haploid in the beginning because the chromosomes are now paired sister chromatids. The total DNA content is diploid, though, because all of the genes are there to make a diploid organism. It is now time to separate these chromatids.

This is what it looks like overall:

Meiosis II

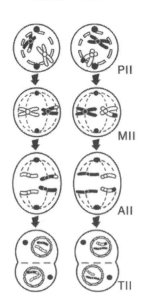

PII

MII

AII

TII

There are more steps to this meiosis II process than are shown here. These twin sister chromatids immediately line up on their own metaphase plate and then divide themselves so that each cell now has just one sister chromatid/chromosome. The amount of DNA is now haploid, or half as much as is necessary to make a whole organism.

A lot of weird things can happen in meiosis that don't happen so much in mitosis. One of these is called *nondisjunction*. It happens when the sister chromatids do not separate equally. The end result is one gamete that is missing a chromosome and one that gets an extra. If this happens to an autosome chromosome, a gamete missing any one of the chromosomes doesn't survive. In a few situations, though, the gamete getting the extra chromosome does survive.

Fun Factoid: What do Down syndrome, Edward syndrome, and Patau syndrome have in common? Each of these is a relatively common trisomy disease compared to other trisomy possible states. In these diseases, chromosome 21, chromosome 18, and chromosome 13 (respectively) are in threes in each cell instead of two (because of nondisjunction). Of these three disorders, only one is survivable: Down Syndrome. Babies born with the other two diseases rarely survive for long.

Failure of segregation or separation of chromosomes in meiosis is much more common in females than it is in males. About 10 percent of meiosis in females goes wrong in this way, especially if the woman is older. These types of failures explain the majority of miscarriages in the early stages of pregnancy.

Fun Factoid: Too many or not enough sex chromosomes is much more common than is true of autosomal diseases. Examples include the X0 disease called Turner Syndrome, where a girl has just one of two possible X chromosomes, and XXY disease (or Klinefelter Syndrome), where a male has too many X chromosomes. What about XYY disease in men? This is called Jacob's Syndrome. It is more likely to lead to very tall men who have a higher risk of being in prison for reasons that aren't completely clear.

Genetic Recombination Means Unique Offspring

Only identical twins are truly identical genetically. All other offspring from the same parents will have similarities and differences between them. This is because of this recombination event that happens in meiosis I (during prophase). When the sister chromatids line up, some genes cross over or switch places. This leads to a different independent chromosome than what was originally present before recombination took place. It is the main reason why the different siblings in a generation will be genetically unique.

Just how unique will these offspring be? It is believed that in humans, one person will be able to make a least 84,000,000 genetically unique offspring based on the mixing and matching of one's mother's and father's chromosomes in meiosis I (two pairs to the 23rd power). It gets more complicated than this because of recombination in prophase of meiosis I. About two to three crossovers happen on each chromosome pair during this time, so really, the chances of identical offspring who are not identical twins are astronomical.

If an organism had just three chromosomes and no recombination occurred, these are the gametes you'd get:

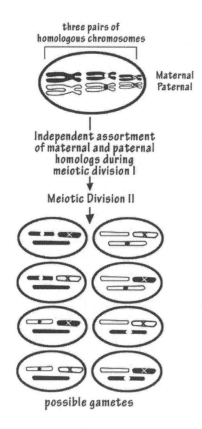

This is what happens to just one of these chromosomes in the recombination process:

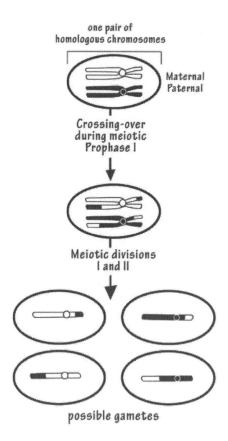

You can see just how complicated it can get and how many possible gametes come from one dividing germ cell in meiosis.

Now, we talked about the fact that one germ cell gives rise to four gametes, but this is only partly true in real life. In humans and most animals, for example, only males will make four gametes, or sperm cells, per episode of meiosis. Females make just one gamete. How does this happen? This is what the *spermatogonium* does in making sperm cells in *spermatogenesis*:

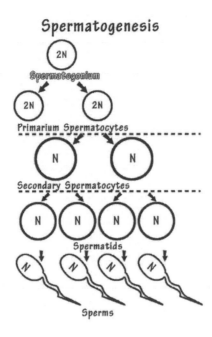

This is what happens instead if the germ cell is an *oogonium* that undergoes *oogenesis*:

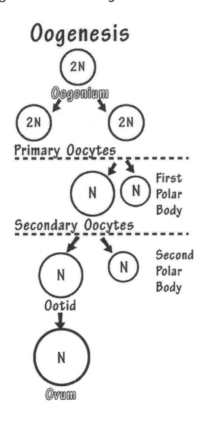

Notice how, at each level, one cell survives while the other is rejected as a *polar body.* This means that three polar bodies get ejected from the system, leaving behind one large ovum and three polar bodies that are dead and get absorbed into the system.

The sex chromosomes do not have recombination happening but they still need to segregate properly. Females with two X chromosomes will pair and segregate like the autosomes. Males have an X and Y chromosome, so these are not at all homologous. These must still pair and divide, however. The only reason this happens is that there are enough base pairs on these chromosomes for them to recognize one another and pair up, even though they aren't homologous.

After meiosis I, X sister chromatids in males separate together and the Y sister chromatids separate together. The next stage, where the primary spermatocytes are made, yields only XX or YY cells. These then become haploid cells that are X, X, Y, and Y haploid sperm cells. It looks like this:

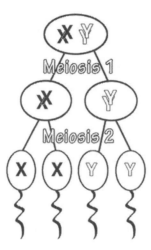

Prophase I, as mentioned, is sometimes a very long stage. In fact, more than 90 percent of all meiosis is in this phase. It is a lot like the G2 phase of an ordinary cell because the nuclear envelope doesn't break down as it otherwise would in regular mitosis. The time it takes to synthesize a new strand of DNA in this phase takes even longer than it does in the S phase of a normal cell cycle.

To Sum Things Up

The process of meiosis only happens in *germ cells*, which are either *spermatogonia* in males or *oogonia* in females. There are two main phases: *meiosis I* and *meiosis II*. In meiosis I, a lot happens to rearrange chromosomes on sister chromatids in order to have unique offspring. The cell divides, with two pairs of *sister chromatids* dividing together. What this looks like is that the amount of DNA is the same as a diploid cell but with a haploid number of separate chromosomal structures.

The second phase is *meiosis II*. DNA never divides and the two similar sister chromatids separate. This gives rise to two cells for each of the cells made in meiosis I that are completely haploid in amount of DNA and number of chromosomes. In males, this gives rise to four gamete cells, or sperm cells, but in females, the end result is just one oocyte, or egg cell. Three of the possible egg cells in meiosis are destroyed as polar bodies.

CHAPTER 33:
CELL COMMUNITIES

Cells rarely live by themselves — even single-celled organisms like bacteria. In this section, we will talk about cells in groups, or communities. In prokaryotes, we are mainly talking about things like *biofilms*, which are communities of bacteria that thrive because they work together. In multicellular eukaryotes like animals, the cells in similar communities are called *tissues*. These also tend to work together through cell-to-cell communication strategies.

Cell Communities in Single-Celled Organisms

Cell communities in single-celled organisms are often called biofilms. These are collections of single-celled organisms that can be made from one species or several. Bacteria, protists, and fungi can coexist in the same biofilm. Some are microscopic while others can be seen by the naked eye, such as dental plaque and pond scum. Biofilms have been found in all places on Earth and in our bodies. Not all of these are dangerous.

All biofilms are wet at least some of the time and many are slimy. They form when freely living microorganisms start coalescing into one place to loosely aggregate or connect with one another. Their main connecting substance, or "goo," is called *EPS*, or *extracellular polymeric substance*. This EPS is made from proteins, sugars, and nucleic acids that allow these organisms to stick together.

This is the basic lifespan of a biofilm from the planktonic state (individual organisms) to the biofilm state (organisms in a community).

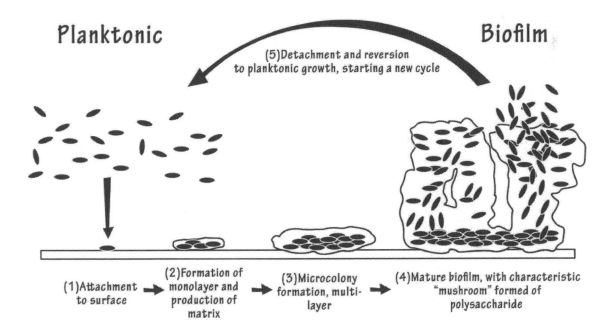

Once these organisms attach, they grow as a group. They start to layer themselves into 3D shapes and complex structures. Channels form between groups of organisms to allow nutrients to flow from place to place (and to allow waste products to be removed). This acts like a primitive circulatory system for these biofilms so that the organisms buried inside have the same chance to get nutrients compared to the outside organisms.

Why Make a Biofilm Anyway?

Biofilms are an example of "safety in numbers." The chances of organisms surviving together as a biofilm are much greater than the chances of each organism by itself surviving a stressful environment. Things like toxins, antibiotics, drying, or changes in pH can make the environment hostile to organisms. The slime, or EPS, can protect the colony and prevent drying out of the biofilm. UV light can be hostile to biofilms but the slime will help the organisms survive better. The slime can sometimes even deactivate things like antibiotics and other toxic substances to the organisms.

Some organisms in a biofilm have the luxury of being dormant. This means that if toxins or antibiotics penetrate the biofilm, those organisms that are dormant are less susceptible to injury. Other bacteria, called *persisters*, don't divide but aren't dormant either. They also resist many types of antibiotics.

Another advantage to being in a biofilm happens if some organisms are *photosynthetic* while others aren't. The photosynthetic organisms can gather light to make nutrients for the heterotrophs that just eat food that is already made. They are symbiotic with one another and help each other out.

Organisms in a biofilm don't have a brain and can't think. The process is natural as long as the organisms can live with one another and the environmental conditions are adequate. If a biofilm is established, it is always possible to have a piece break off and start its own biofilm. These cells can still communicate with each other in interesting ways.

One major way of communicating in a biofilm is called *quorum sensing*. This is a way that cells communicate with one another both near and far. If the messages involved in quorum sensing say one thing, the group acts as a whole, doing something they've all agreed on. In human infections, quorum sensing can say if a few bacteria will break off to start an active infection, for example.

So, biofilms are like small cities that live and work as a group. They may or may not all be the same, but all contribute in some way to the total functioning of the biofilm. A few organisms will have flagella to move around, jockeying for the right position before finally settling in one place. Once the goo gets created, the organisms adhere to this stickiness and to each other.

All humans have biofilms in their bodies. Some are common — like dental plaque, which is basically a biofilm that the dentist has to scrape off. Others are uncommon and settle around prosthetic joints, catheters, pacemakers, or mechanical heart valves. Even contact lenses will eventually develop a biofilm in your eyes. Once they are established, they are really hard to get rid of. They are the hardest to get rid of if they are attached to something non-living in your body.

Your Body's Communities Are Tissues

In biological systems where the organism is multicellular, communities of cells of the same type are called *tissues.* Tissues are held together by their own kind of glue called the *extracellular matrix.* The different types of tissues in the body each have their own special function.

With so many types of organs in the body, you'd think that there would be dozens and dozens of types of tissues. In reality, there are only four main types. Within these four types, however, the tissues can look different depending on where in the body they are located. For example, fatty tissue and blood are of the same type of tissue but they look nothing like each other.

Let's talk about the four types of tissues seen in animals. These are *connective tissue, muscle tissue, nervous tissue,* and *epithelial tissue.* Here's what they look like:

Four types of tissue

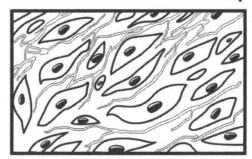
Connective tissue

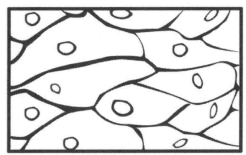
Epithelial tissue

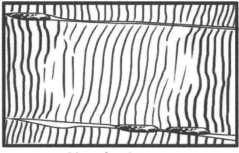
Muscle tissue

Nervous tissue

Epithelial Tissue

Epithelial tissue is also called *epithelium*. Most surfaces in the body are lined with this type of tissue. Your skin, the lining of your airways, the lining of your gut, and the inside of glands and even blood cells are all made of certain types of epithelial tissue.

These tissues are highly connected to one another, so they can form a barrier to things getting into the body. They often have tight junctions between the cells in order to create these tight barriers and often have cells buried in them to produce mucus or other secretions. Other cells in these tissues will absorb things from the outside.

Epithelial Tissue

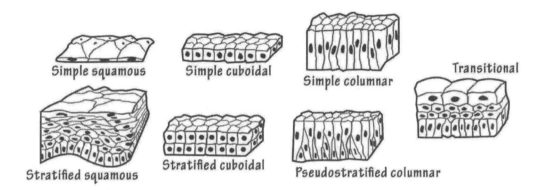

These different epithelial layers have cells of varying shapes and layering capabilities.

What do epithelial cells do? Here are some of their activities:

- To cover or line cavities or surfaces of an organ
- To protect the body from the environment (skin mainly)
- To absorb nutrients and water
- To get rid of waste products
- To secrete certain hormones, mucus, saliva, sweat or enzymes in glands and related tissues

In general, layered epithelial tissue means that stress is on the tissue somehow, especially if the tissue has a protein like *keratin* to protect it further from moisture. Keratin is what prevents moisture from leaving or entering the body too freely. Other epithelial tissues are major hosts to mucus-making cells, called *goblet cells*. These are found in the gut and the respiratory tract.

Connective Tissue

Connective tissues make up a lot of the body. These are clusters of related cells that have a little or a lot of extracellular matrix (nonliving proteins and carbohydrates, mainly) between the cells. The extracellular matrix can be completely tough, such as the kind you'll find in bone (which is a type of connective tissue). Blood, too, is a connective tissue, but the matrix is completely liquid.

The three main types of connective tissue are *fluid-based, skeletal,* and *fibrous.* Ligaments, bones, fatty tissue, and the spaces in between organs are often filled with connective tissue. These are the main types of connective tissue you'll see in the human body:

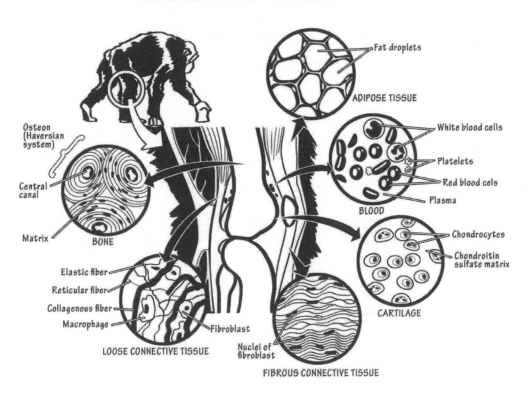

Muscle Tissue

When you think of muscle tissue, you probably think of the muscles that move your body, right? This is definitely muscle tissue but there are other types. Your heart has *cardiac muscle tissue* and your blood vessels, GI tract, and bronchial tree all have *smooth muscle* in them. The main feature of muscle tissue of all types is that it will contract in some way. They get a signal from nerve cells and then contract.

Cardiac muscle is unique because the signal to contract can start with one cell and will spread from cell to cell directly through what's called a *muscle syncytium*. Heart muscle cells also do not necessarily have an outside signal, which explains why a heart surgeon can take out a beating heart and it will still beat for a while!

This is what the different muscle types look like:

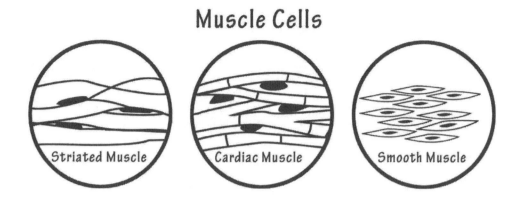

Muscle Cells

Striated Muscle

Cardiac Muscle

Smooth Muscle

Nerve Tissue

This is also called *nervous tissue.* There are different cell types in nerve tissue. Some are the nerve cells themselves. These are electrically active cells that can take a signal from the sense organs or from another nerve cell, causing it to depolarize or change the electrical charge across the cell membrane. This signal gets sent along the nerve cell, which can be as long as three meters in total length, to cause an effect on another cell. Usually, the target cell for these messages is either part of other nerves, muscles, or glands.

There are supporting cells in nerve tissue as well, called *glial cells*. There are many types of glial cells in nerve tissue. Some just provide structure or nutritional support. Others make *myelin*, which coats some nerve fibers so the signal happens faster. Still others act as brain *immune cells* to fight off any possible infection that could get into the brain. This is what nerve tissue looks like:

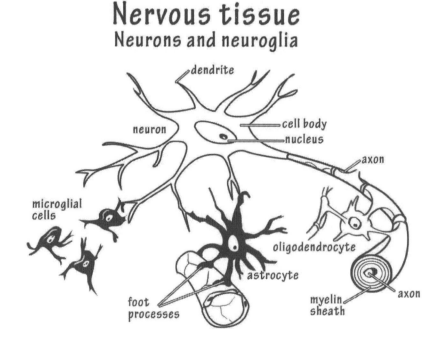

Nervous tissue
Neurons and neuroglia

dendrite

neuron

cell body

nucleus

axon

microglial cells

oligodendrocyte

astrocyte

foot processes

myelin sheath

axon

Notice the nerve cell plus all of the different supporting cells, including *microglia* (immune cells), *astrocytes* (nutrient cells), and *oligodendrocytes* (myelin-making cells).

To Sum Things Up

Cells can live in different types of communities. Microbes like bacteria, fungi, and protozoa will come together to make communities called *biofilms*. They can get very large and are held together by goo called *EPS*. Once they are held together, they often act as a unit to help the whole community survive harsh environments.

The cells in multicellular animals form communities called *tissues*. The tissue types in humans are *epithelial, connective, muscle,* and *nervous.* They look and act differently. Most organs of the body are made from more than one community or type of tissue.

SECTION TEN:
GENE EXPRESSION MAKES EACH CELL UNIQUE

This section brings a lot of things you've learned full circle. Every cell in the body that has a nucleus participates in one major goal: to take their genetic material, use the parts of it that are needed, and make proteins. They need to do this with the minimum number of errors and make sure that it's done efficiently. In other words, would you be an efficient cell if you made every protein the genome coded for? Certainly not, largely because every cell is unique and is most efficient when it makes just the proteins it needs at any given moment.

CHAPTER 34:
CENTRAL DOGMA

The *central dogma* is a fancy way of saying "the big picture." This is really what cells are made for, which is to survive by taking their own genetic material (their DNA genome), create a unique message for required proteins, and turn that message into a workable protein for the cells (or a substance the cells export elsewhere). It's very simple in theory but involves a great many steps.

There is a trifecta of things in the central dogma you should remember. These include the following:

- **Replication** — This is when the cell's genome in its entirety gets copied in order for DNA to get passed onto the next generation of cells.
- **Transcription** — This is the DNA-to-RNA process. It's a shift in the message. It's like the genes are shut-ins that can't leave the nucleus. To make up for this, these genes transcribe a message to a *messenger* (usually messenger RNA or mRNA). This messenger can find its way out of the nucleus in order to get it to the ribosomes for protein synthesis.
- **Translation** — This is the most unique part of the central dogma. It involves the translation of a message in one type of biomolecule (a nucleic acid) to another (a protein). It always takes place in some type of ribosome.

This whole process is irreversible and is what causes a cell to survive on a regular basis and to ensure the survival of its lineage. These are the basic steps in visual form:

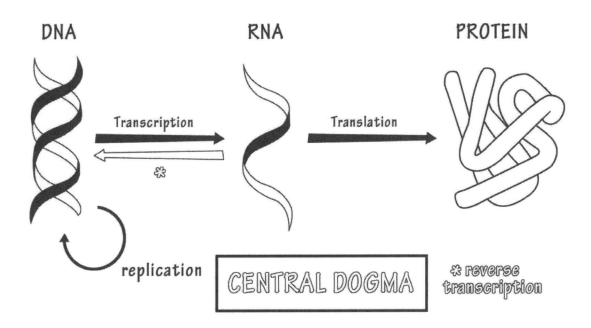

What is the purpose of a *dogma*? Basically, it's helpful simply because it is the "big picture." You should keep this entire picture at all times when you think about nucleic acids and their function as well as proteins and their production.

There are three categories of transfer possible in the central dogma. These are *general, special*, and *unknown* transfer categories. Let's look at these:

General Transfers

There is basically one set of three transfers that happen in all *nucleated cells*. There are three parts of the general transfer process:

- DNA replicates itself into an exact copy of its own genome. To refresh you on this process, see this image:

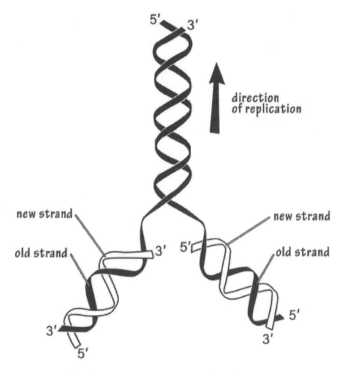

- DNA information gets transcribed into an *mRNA molecule*. Here's what this looks like:

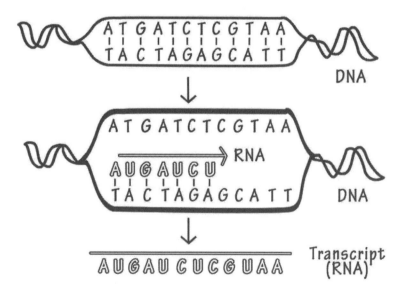

- The mRNA molecule gets translated into a protein.

Special Transfers

These are transfers that are important but not the main purpose of the cell. These include the following processes:

- **RNA replication** — This is when there is a piece of RNA that gets copied into another piece of RNA. It happens a lot in certain viruses and looks like this:

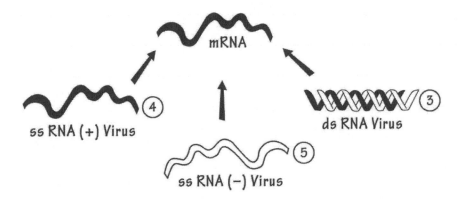

- **Reverse transcription** — This is when RNA gets copied into piece of DNA. This involves the enzyme called *reverse transcriptase*. It is extremely important for certain viruses, such as HIV. This is what it looks like:

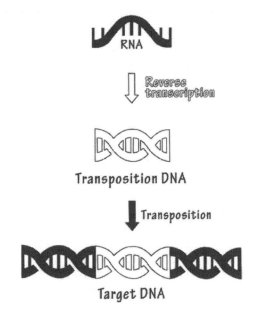

- **Proteins get made from DNA directly** — This is when a protein gets made but there is no RNA intermediate.

All of this essentially means that the central dogma is not perfect and is not the whole picture of what happens in a cell.

Unknown Transfers

There are three of these so-called unknown transfers because they are not known to actually occur in nature. These include the following:

- A protein gets copied from another protein.
- RNA gets made using a protein as a template.
- DNA gets made using a protein as a template.

So, that's basically the central dogma. Before we go on, let's look at some fundamental differences between *replication* (which we've already discussed) and *transcription* (the making of mRNA from a strand of DNA). Both involve the transfer of nucleic acids. These are the major differences:

DNA Replication	RNA Transcription
Replication involves taking the genome and copying into another DNA genome.	Transcription involves the taking of sections of the genome (genes) in order to make RNA.
The most important part of this is that it allows the transfer of the genome to the next generation.	The most important part of this is that it allows genes to be regulated and then to make an RNA message molecule for protein synthesis.
DNA goes to make identical new DNA molecules as its sole purpose.	DNA goes to make an RNA molecule (usually messenger RNA) in specific segments only.
It happens specifically in the S phase of the cell cycle for the purposes of genetic preservation.	It happens specifically in G1 and G2 in the cell cycle but gets suppressed in the S phase.
It prepares the cell for cell division to occur.	It prepares a nucleic acid message in order to set up the process of translation to occur in the ribosomes later.
It is exclusively involved in the process of cell division.	It must involve some aspect of gene regulation before the process can evolve further.
The raw materials are dATP, dTTP, dGTP, and dCTP.	The raw materials are all of the nucleotides like ATP, GTP, UTP, and CTP.
The template for this process is both DNA strands at the same time.	The template for this is a segment of single-stranded DNA.
The process involves a necessary RNA primer in order for it to work.	The process involves an immediate primer in order to have it work.
The needed enzymes include DNA helicase and DNA polymerase, among others.	It involves transcriptase (a kind of DNA helicase) and RNA polymerase.
It is necessary to unwind and split the entire DNA molecule at some point.	It involves the unwinding and breaking up of certain segments of DNA at a time.
The base pairing that occurs involves adenine with thymine and guanine with cytosine.	The base pairing that occurs involves adenine with uracil and guanine with cytosine.

The entire template strand gets copied as the process proceeds.	Only a segment of the DNA template gets a complementary strand but each strand of DNA could be a potential template for something.
The end result is the formation of two segments of daughter DNA genomes.	The end products are tRNA, mRNA, rRNA and other minor RNA types.
The type of strands produced will be double strands of DNA.	The type of strands made are always single-stranded pieces of RNA.
The joining of Okazaki fragments on the lagging strands completes the process.	Editing of the messenger RNA completes the process.
The DNA end product does not need any specific processing.	The molecule made (the RNA transcript) needs to be extensively processed in order for it to be fully functional.
The DNA created is bonded through hydrogen bonding to the original DNA template strand.	The transcribed piece of RNA will separate from the template it was created from.
The product does not really leave the nucleus except when it disappears in mitosis.	Most of the product made leaves the nucleus and enters the cytoplasm.
The products made are never degraded later.	Once the function of the end product is completed, the product is later degraded.
This happens at a rapid rate that is about 20 times faster than the process of transcription.	The process happens much more slowly than in replication.
The next steps are either transcription or another replication in the daughter cell.	The next step after this is translation, especially for messenger RNA.

So, that's the entire purpose of being a cell wrapped up in a neat process called the *central dogma*.

To Sum Things Up

The *central dogma* involves the three main cell processes of an actively dividing cell. First, it must replicate its genome (the entirety of its DNA) in order to pass this onto the next generation. Second, it must use the genome in significant segments in order to pass the DNA message to a messenger RNA molecule that can leave the nucleus. Finally, the mRNA message gets translated into a protein molecule.

There is one thing we now know happens that is not a part of the central dogma. This is called *post-translational modification of protein.* It mostly happens in the endoplasmic reticulum and involves various chemical additions and subtractions to the protein molecule after it has been translated in the ribosomes.

CHAPTER 35:
GENES MAKE PROTEINS THROUGH THIS PROCESS

In this chapter, we will talk about the process by which a gene (a relevant section of DNA) gets transcribed into a section of mRNA (called the *primary transcript*) and how this transcript gets modified if necessary and finally makes a protein in the ribosomes. This process is similar in prokaryotes and eukaryotes but these are different enough to talk about them separately.

Prokaryotic Transcription

Prokaryotes have no nucleus, so the transcription of DNA to RNA is almost always paired with translation. It is a lot like linework in an automobile factory. This means that the strand of DNA is transcribed while the piece of RNA is getting translated into a polypeptide chain in the bacterial ribosomes. This is what it looks like:

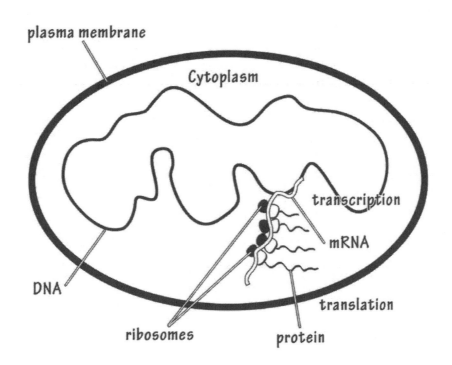

As is true of anything, there must be a beginning, middle, and end. With transcription in prokaryotes, the official names for these things are *initiation, elongation,* and *termination.* Let's start with the major enzyme involved.

Bacteria have just one type of enzyme for the making of the RNA transcript. This is called *RNA polymerase*. It's a messy enzyme with six separate subunits called α, β, β, 'ω, σ (there are two alpha subunits). The whole enzyme is called the *holoenzyme*. It looks roughly like this:

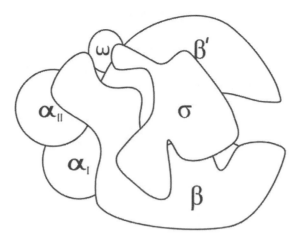

The first step is initiation, or starting the process. There is a specific site along the circular piece of bacterial DNA called the *promoter site.* Just remember that it "promotes" the transcription process. Once the enzyme attaches, the DNA double helix starts to unwind.

As this enzyme binds to the promoter site (which is just 40 to 60 base pairs in length), it starts to build a short section of messenger RNA. There is a specific sequence of DNA in this promoter region that is so important that you'll find it in almost all bacterial species. This is called the *TTGACA segment* (because of the base pairs involved). It's where the *sigma factor* (or the σ factor) interacts with the DNA to be transcribed and starts the entire process. This TTGACA box is called the *Pribnow box.* Unlike *DNA polymerase*, this *RNA polymerase* enzyme doesn't need primer.

This is a closeup view of the promoter region in bacterial DNA:

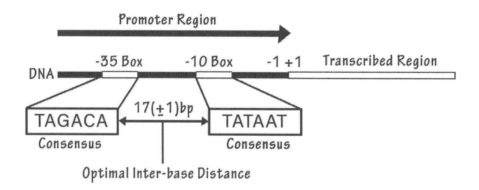

Next comes the *elongation phase.* The sigma factor drops off, so it is no longer called the *holoenzyme*; instead, it is called the *core enzyme* and is made of the other subunits (α2 ββω). This is the part that moves on to complete the elongation process. The beta subunit is the main part that continues the elongation process. Remember that all nucleic acids are made in the 5' to 3' direction and need the four ribonucleoside-5-triphosphates as precursors.

The DNA template, RNA polymerase, and the new piece of RNA is called the *ternary complex.* The entire piece of DNA that is unwound and separated is called the *transcription bubble*. This is the whole process of prokaryotic transcription, including this bubble:

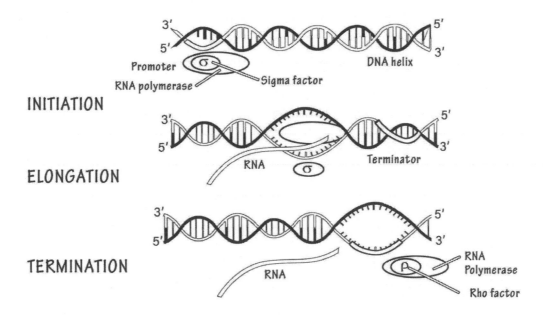

INITIATION

ELONGATION

TERMINATION

Prokaryotes save space by using both strands of DNA but not necessarily at the same time. Depending on the cell needs, the organism will use whatever strand makes the protein it wants. The conditions of the cell determine where the RNA polymerase attaches at any point in time.

The final stage of transcription is *termination.* Just as there is a segment of DNA that signals the "start button" of transcription, there is a "stop button" too. Most commonly, it will be series of GC matches followed by a series of AT matches. This often causes the last segment of RNA made to fold back on itself to make a hairpin-like structure. If this doesn't happen, a special protein called the *rho protein* is needed to cause the process to stop and for the RNA to separate away from the DNA strand.

Prokaryotic Translation

As we already talked about, protein translation usually happens at the same time as transcription. It involves ribosomes in the cytoplasm of the cell that are busily making proteins whenever they come in contact with a piece of mRNA. (Remember that there is no endoplasmic reticulum in these cells.) This is a simple version of what it looks like:

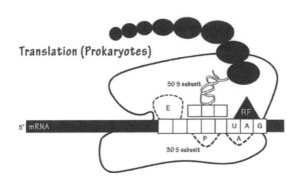

The bacterial ribosomes are made from two separate subunits of ribosomal RNA and protein. When the two parts come together, they create a larger unit called the *70S unit*. The messenger RNA runs through this structure so that this can all turn into one big protein-making factory. Like everything with nucleotides, things are read and processed in a 5' to 3' direction.

Ribosomes are really impressive. There are three main sites, called binding sites, that are important in these structures and that help the ribosomes function. These include:

- **A Site** — This is also called the *aminoacyl tRNA binding site.* This is where the "loaded" tRNA first attaches. A loaded tRNA molecule is called an *aminoacyl-tRNA* because it has the amino acid necessary to add onto the growing polypeptide chain.
- **P Site** — This is also called the *peptidyl-tRNA binding site* because it is where the bound and loaded tRNA acts to get into the vicinity of the peptide chain.
- **E Site** — This is the site where the tRNA binds after dropping off its amino acid load to the growing peptide chain and from where the empty tRNA exits.

This is what the ribosome and its binding sites look like:

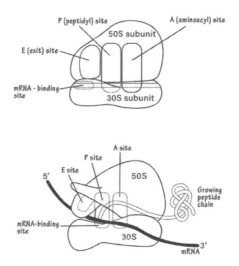

This is what the whole process looks like:

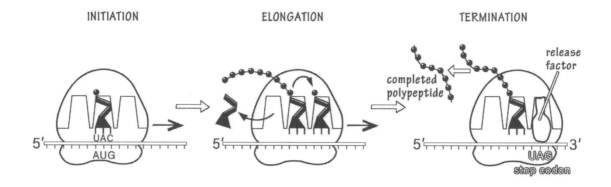

The initiation phase happens when mRNA and the ribosome come together to make a *complex*. The first codon, called the "start" codon, is almost always *AUG (adenine-uracil-guanine)*. The first aminoacyl tRNA attaches, which is called the *initiator transfer RNA*. The elongation phase is a gradual elongation of the peptide chain — amino acid by amino acid — until the stop codon is achieved. There is no matching tRNA to the stop codon, so the polypeptide chain drops off.

There are as many different tRNA molecules as there are amino acids (and then some). Each tRNA molecule binds to the associated amino acid it was meant for with a *covalent bond* between the two. It's not actually called "loading" the tRNA but is instead called *amino acid activation*. There is just one amino acid per transfer RNA. The enzyme that does this is called *aminoacyl-tRNA synthetase*. There is actually a separate synthetase enzyme for every combination of amino acid and tRNA.

Fun Factoid: If every protein synthesis event starts with the same start codon, does this mean that every protein on Earth starts with the same amino acid? Yes and no. The AUG start codon almost always codes for the amino acid methionine, so this is the start of every polypeptide made on Earth. But the difference is that some proteins will later get modified to remove this amino acid before its final structure is completely made to be used by the cell.

There are three stop codons on an mRNA molecule possible. These will be *UAA, UAG,* and *UGA.* Prokaryotes have no matching aminoacyl-tRNAs to match these codons. Instead, there are release factors that bind to these codons. There is RF-1, which binds to UAA and UAG; there is also RF-2, which binds to UAA and UGA. There is a third one, known as *RF-3*, that helps the two others do the actual binding to the ribosome.

What happens when the release factors bind and interact with the ribosome and mRNA? There is no peptide chain formation at all. Instead, the end of the polypeptide chain is prompted to bind instead to water, which essentially "caps off" the end of the peptide chain. This allows the free polypeptide to leave the ribosome completely.

Transcription in Eukaryotic Cells

Eukaryotes have some differences in how transcription occurs, even though many parts are very similar between the two types of organisms. Of course, among us "higher" organisms, the process is a bit more complicated than it is with prokaryotes. There are, for example, many transcription factors in eukaryotes that bind to the promoter region for the gene, which allows the RNA polymerase to also attach to the area it needs.

There are three different RNA polymerases, which makes things more complicated, too. Each of the polymerases are themselves bigger and more complex as well — all of them are made from at least 10 different subunits.

RNA polymerase I is actually found in the nucleolus. This is because it is used only for the making of ribosomal RNA (which is made inside the nucleolus). The ribosomal RNA made by this enzyme is not really considered the same as regular nucleic acid because the end result is a molecule that is mostly structural. No proteins are made from these RNA molecules.

RNA polymerase II is found in the nucleus. It codes for all of the messenger RNA in the cell. Messenger RNA isn't what gets made. Instead, it's *pre-mRNA*. It's called this because in eukaryotes, there is a lot of processing that goes on to make a messenger RNA "ready" to leave the nucleus and make proteins in the cytoplasm.

The third RNA polymerase is *RNA polymerase III*. This is also found in the nucleus and makes all of the tRNAs and small nuclear RNAs. One of the ribosomal RNAs (called *5S rRNA*) is also made with this particular enzyme.

Eukaryotes have a coding sequence that indicates that the gene transcription is supposed to begin. This is called the *TATA box*, although the real sequence is *TATAAA*. It isn't immediately next to where the transcription starts, or "initiates," but is about 25 to 35 bases away.

There is also no sigma factor in eukaryotic RNA polymerase. Instead, there are separate transcription factors. There are several called *TFII factors* or *basal transcription factors* whose job it is to act as molecules need to help promote gene transcription. Some of these add together to make a giant complex that says, "Start transcribing this gene," and tells RNA polymerase II to bind to the proper spot. This is what it might look like:

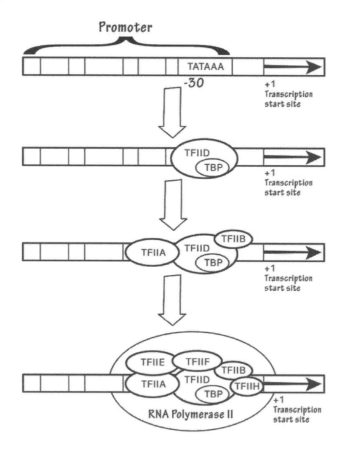

There are other transcription factors in eukaryotes that bind to various parts of the chromosome as well. Some are called *enhancers*, while others are called *silencers*. Some of these are found a large distance away from the actual genes they enhance or silence. See how it gets more complicated in eukaryotic cells? It turns out that these enhancers and silencers are a nice thing to have to make transcription more efficient, but really aren't absolutely necessary for transcription to happen at all.

When transcription proceeds past the initiation point, the RNA polymerase detaches and goes on its own to make the RNA transcript. Just like all nucleic acid synthesis, the enzyme chooses which bases to add to the new transcript by reading the sequence of bases on the DNA segment. It makes a matching strands of RNA by having A added when it reads a T on the DNA strand, a G added when it reads a C on the DNA strand, and so on.

Fun Factoid: How does a cell regulate which genes to make? If you know the whole answer to that, you'd probably win the Nobel Prize. One interesting way we know of relates to the fact that DNA in eukaryotes is packed with histone proteins that help condense it. A gene won't be able to unwind or transcribe if it is too wrapped up in a glob of histone proteins.

There is a special protein called the *FACT protein*, which basically means *facilitates chromatin transcription*. Its job is to move histone proteins out of the way in order for the RNA polymerase to transcribe the gene. It's this same FACT protein that puts the histone proteins back in place after transcription is finished.

Prokaryotes make relatively intact mRNA molecules; this isn't the case, though, for eukaryotes. The mRNA molecule made in eukaryotes is at least 1,000 or more bases longer than it needs to be at the outset. This extra section of mRNA is called the *pre-mRNA tail.* Other RNA types don't have this long tail. The RNA polymerases I and III both have a specific sequence of DNA they read, which says to stop transcribing at that point.

mRNA Needs Processing

For reasons we don't know enough about, pre-mRNA must be processed a lot before it can do its job. There are parts that need to be cut out from the middle of the mRNA strand. These are called *introns*. Oddly, it is the *exons* that are kept as part of the mRNA that gets translated. (Hint: Think "exit" to remember "exons" because they get to exit the nucleus.)

There are other changes that happen as well that help to make the mRNA less likely to degrade or fall apart before exiting the nucleus. These are called the *5' cap* and the *poly-A tail.* This image shows you what the final product looks like. The whole process is called *RNA editing* or "splicing of RNA":

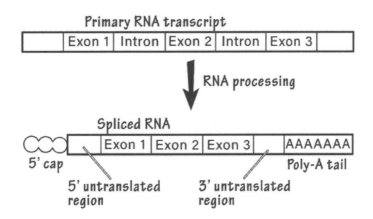

This 5' cap added to the 5' end of mRNA is actually a side chain of *7-methylguanosine.* This is helpful to the mRNA molecule because it helps the ribosome "know" where to start the translation process. The poly-A tail is a long tail of about 200 adenosine bases. Its job is to keep the mRNA from being degraded before it leaves the nucleus.

The whole role of spicing is not very clear yet. The pre-mRNA made is much longer than necessary and contains big sections that never turn into protein in translation. These introns must be cut or spliced out in some way. The theory is that the introns are important to the regulation of genes but aren't necessary later. You can imagine that it would be very important to splice out these introns at the *exact* bases where they start and stop. If this doesn't happen, the whole protein made by what's left over would be messed up.

These introns are marked with a GU sequence at one end and an AG sequence at the other. This is partly how the cell knows where to cut the introns out. There are huge complexes called *spliceosomes* that cut introns out in mRNA splicing. Most of the spliceosome is made of small nuclear RNAs and proteins. They look like this:

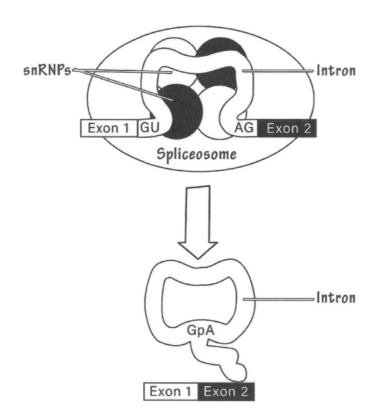

See from this image how the intron gets cut out, leaving just the exons attached to one another? This is how the mRNA gets modified in order to be just the right molecule for later translation in the ribosomes. Medical researchers have found that cancers and other diseases are caused by problems in splicing, mostly because these can lead to wacky proteins that don't do anything, or do the wrong thing entirely. There are as many as 70 introns in just one section of pre-mRNA.

Eukaryotic Translation: Protein-making Machinery in the Ribosomes

Translation isn't much different in eukaryotes as it is in prokaryotes. The ribosomes are a bit different in size but not at all in function. Remember that ribosomes are a mixture of rRNA and proteins that are slightly different from organism to organism. The whole process takes a combined effort from the *ribosomes, tRNA, loaded amino acids on the tRNA, mRNA, and enzymes* to accomplish this translation process.

Each cell has thousands of ribosomes per cell. These are mostly clustered on the rough endoplasmic reticulum. The two halves are the *40S subunit* and the *60S subunit.* These are bigger than prokaryotic ribosome subunits. The smaller one helps to attach the mRNA molecule, while the larger one helps to attach the tRNAs to the ribosome complex. The entire complex together is not called a ribosome anymore but is called a *polysome.*

When the tRNA molecule gets loaded, or "charged," with an amino acid, the process requires *ATP energy*. The main reason for this is that the cell wants to make a high-energy bond between the amino acid and the tRNA molecule. It uses this extra energy to allow the amino acid to attach to the polypeptide chain in the ribosome without needing any extra energy. In a sense, the "energy" is put into the bond itself before the tRNA does its job.

Because the processes of initiation, elongation, and termination of the polypeptide chain in the ribosome is not much different between the prokaryotes and eukaryotes, we won't elaborate on this further. The main difference you'll see in eukaryotes, however, is in the modification of the protein after translation has happened. This is called *post-translational protein modification*, which doesn't happen much in prokaryotes.

What does this modification process look like? It involves a lot of things, in fact. Here are a few:

- **Protein folding** — This generally happens naturally and is based on which amino acids are hydrophobic, hydrophilic, or have attractions like acid-based attractions or hydrogen bonding between the amino acids. There are *chaperone proteins* that help in some cases.
- **Cleavage** — As mentioned before, often the methionine amino acid at the beginning of the polypeptide chain gets cleaved off; however, other amino acid segments can also get removed in order to make the final protein molecule.
- **Side chain addition** — This happens in the endoplasmic reticulum and involves the addition of things like sugars, methyl groups, other carbon-based groups, or a molecule called *ubiquitin* to the protein.
- **Signal sequence addition** — This is the sequence of amino acids that helps the cell know where a specific protein is supposed to go. These get tacked to the front end of the amino acid so that the cell gets the right message as to where it belongs by looking at this signaling segment.
- **Other modifications** — These involve other things like covalent bonding between amino acid residues. For example, cysteine has a *sulfhydryl group* that gets combined with another cysteine residue to make the disulfide bridge between the two amino acids.

Ubiquitination

Ubiquitination means adding a molecule of *ubiquitin* onto the peptide chain. Generally, this is not good for the targeted protein because it often spells their doom. Ubiquitin is a small protein that is found in the endoplasmic reticulum; it is a major part of what's called the *ubiquitination proteasome pathway*. Here's how it works:

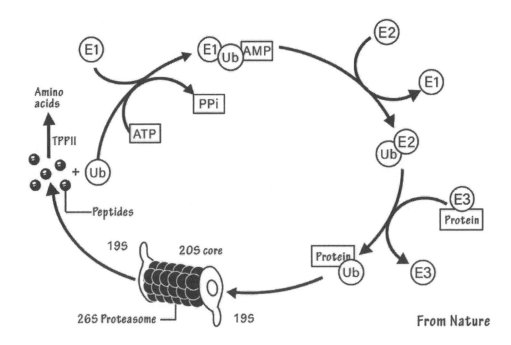

Let's say that a protein is misfolded or "wrong" in some way. The cell recognizes this and starts this process of getting rid of it. First, the ubiquitin must be charged with the E1 enzyme and then the E2 enzyme (as shown in the pathway). This requires ATP energy. The misfolded protein is marked with another enzyme, called the *E3 enzyme*. This allows the misfolded protein to be marked with the ubiquitin protein itself.

Then, the protein is further targeted by a "tube-shaped" structure called the *26S proteasome,* located in the membrane of the endoplasmic reticulum. The protein that enters this tube does not come out intact but is degraded completely to make amino acids again. The amino acids get recycled, of course — hopefully in order to make the right protein the next time around.

To Sum Things Up

The processes of *transcription* and *translation* happen slightly differently in prokaryotes and eukaryotes. Prokaryotes often perform the transcription process and translation process at the same time. This means that DNA goes to RNA and then to protein — all in the cytoplasm of the cell. The process involves just one major enzyme, called *RNA polymerase*. We will talk more about how the prokaryotic genes are regulated later.

Transcription in eukaryotes is more complex than it is in prokaryotes. This happens in the nucleus or nucleolus of the cell. There are three separate RNA polymerases and a number of transcription factors that are necessary to have this process of turning DNA into RNA happen.

The *messenger RNA* that gets made after transcription must be edited before it can leave the nucleus. There are things like *5' endcap addition* and the addition of the *poly-A tail* that will help the mRNA resist being degraded before leaving the nucleus. The pre-mRNA initially made must get spliced as well. This involves the removal of *introns* from the molecule so that only the exons get "expressed" as protein later.

Translation in both prokaryotes and eukaryotes happens in the ribosomes. It is basically the same for both types of organisms. The process involves the addition of amino acids onto a growing polypeptide chain in the ribosome after the mRNA message and the loaded tRNA molecules get together to help this happen.

After a polypeptide chain is made, it is usually modified somehow in eukaryotes. A number of processes happen to make a protein "complete," such as *folding* and *side-chain addition.* The end result is that the right proteins go on to where they are needed, while the misfolded proteins get degraded in the endoplasmic reticulum.

CHAPTER 36:
DNA REPAIR AND RECOMBINATION

In this chapter, we will talk about how DNA is repaired — both at the time it is made and at other times. There are some errors that happen during DNA synthesis and those that can happen when DNA is damaged or *mutated*. We will discuss how these are fixed. We will also talk about how *recombination* works in the process of meiosis.

DNA Repair

If you thought that DNA replication was perfect every time it happened, you would be wrong. The molecule that does this (*DNA polymerase*) makes mistakes all the time. Even so, the error rate is low for this enzyme: about one mistake per 100 million to one billion base pairs. This is really necessary because you wouldn't have a successful passage of a cell's genome to the next generation without a good repair mechanism. The species would not survive and, in multicellular organisms, there would be so many mutations that could lead to constant cancer formation.

How do DNA mistakes happen? The most common mistake is that the wrong base gets added as a supposed "complementary" base to the one seen on the template. Thymine can bind with guanine, for example, and cytosine can bind with adenine. This is called a *wobble* and it looks like this:

Thymine-guanine wobble

Cytosine-adenine protonated wobble

Another error that can happen is called *strand slippage*. This happens when a base is missed so that the new strand has a deletion of the complementary base. An extra base can also get inserted into the new strand of DNA. If you can imagine the new strand forming a loop as it is developing, you can see where a larger insertion can also occur. This is what it could look like:

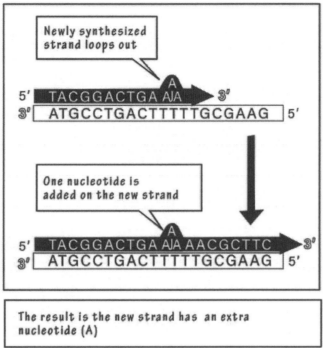

The following labels appear within the figure:

Newly synthesized strand loops out

5' TACGGACTGA AJA 3'
3' ATGCCTGACTTTTTGCGAAG 5'

One nucleotide is added on the new strand

5' TACGGACTGA AJA AACGCTTC 3'
3' ATGCCTGACTTTTTGCGAAG 5'

The result is the new strand has an extra nucleotide (A)

How do these errors get repaired? First of all, it's estimated that if no repair happened at all, a single human cell would make about 120,000 errors every time the cell needed to divide. With repair mechanisms, this doesn't happen often. Proofreading happens using the DNA polymerase molecule itself. As it lays down new nucleotides, it checks itself in a process known as *mismatch repair*. It simply stops when a mistake is made and switches the mismatched nucleotide with the correct one. This works more than 99 percent of the time.

Secondly, if there is an uncorrected error, the DNA molecule made will often be deformed in some way. This is detected by other enzymes that will find the deformities and correct them immediately. This reduces the error rate even further. Those mistakes not corrected are called *mutations.*

Technically, a mutation is any error in the DNA molecule that happens at any time. A molecule of DNA is said to be mutated if the strands (or strand) does not have the intended order or type of bases in its structure (compared to its parent DNA molecule). Not all mutations are bad. In fact, many are neutral. If the DNA has an error that doesn't change the protein made in the end, it will be a *silent mutation*.

The environment can also cause DNA to be mutated. There are many environmental exposures that can mutate the DNA of a cell. Some (but not all) of these can be repaired at some point. It's also possible to have a random spontaneous mutation with no known cause. These spontaneous mutations happen only because a base can be chemically altered for no particular reason. An example is called *depurination*, where the purine bases get detached from the rest of the DNA molecule. Another is *deamination,* where an amino group gets lost from any of the nucleotides.

Fun Factoid: Which mutations would be the worst kind to have? Certain germline mutations (those that happen to reproductive cells) are bad because they get passed to the offspring. Also, any mutation that happens in a gene that is responsible for regulation of cell division (such as the HRAS mutation and the p53 mutation) is bad because these kinds of genes prevent uncontrolled cell growth. If you guessed that this could lead to cancer, you'd be right about that.

These are some things that cause DNA mutations:

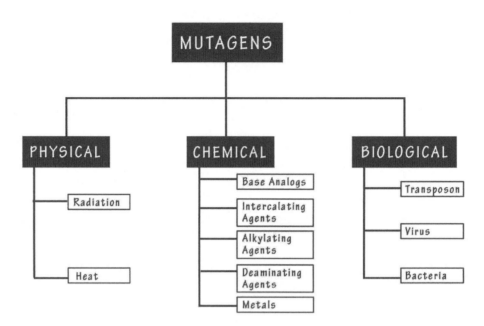

Let's look at an example of how a mutation can happen. Take sunlight, which is a form of radiation that gets into your skin cells but doesn't have the electromagnetic energy to get any further into the body. This UV energy can cause the pyrimidines next to each other to bind together, creating a *pyrimidine dimer*. These changes will bend or distort the DNA. This is what it looks like:

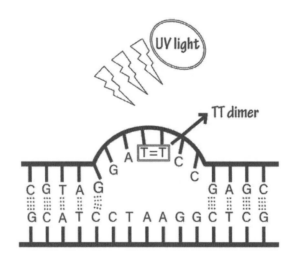

The p53 gene that regulates the cell cycle (and cell division) seems to be very vulnerable to this kind of damage. This is how a lot of cancers, including skin cancer, can develop. It's also why p53 is called an *oncogene*, or cancer gene.

There are repair mechanisms for this type of DNA damage, called *nucleotide excision repair.* This is what it looks like:

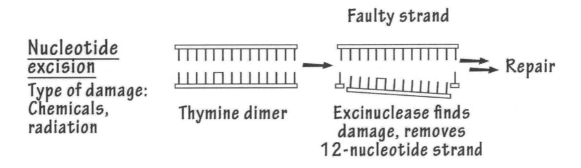

There are enzymes that find these pyrimidine dimers. They remove the dimer, plus a bunch of other bases on either side. Then, they simply fill in the missing space so that the DNA strands are both normal.

Another repair process happens if the wrong base gets added to the DNA strand, leading to a mismatched set of base pairs. There are enzymes that detect the mismatch. They can cut out the wrong base and replace it with the correct one. It looks like this:

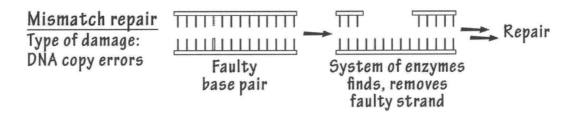

A third type of repair is called a *base excision repair.* This is when the base on a strand of DNA is chemically damaged somehow. Again, there are enzymes that can detect this and simply snip out the damaged base, allowing normal repair mechanisms to put in the right one. It looks like this:

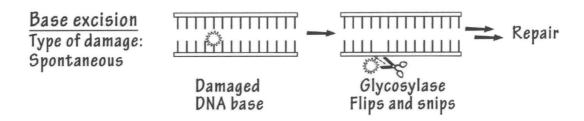

Fun Factoid: There is a human disease called xeroderma pigmentosum, or XP. People with XP are really prone to getting skin cancer. The reason why this happens is that they have a defective enzyme used to repair the typical UV damage to skin cells that happens all the time. If you inherit the disease from

both of your parents, you have no good copies of the gene to make this enzyme, so you can't repair sun-damaged skin. This leads to a high risk of sun damage.

One of the worst kinds of DNA damage happens with exposure to *gamma rays* or x-rays, which are very high in energy. These can cause double-stranded breaks in DNA and lead to chromosome rearrangements, where one piece of a chromosome tacks onto the end of another. It can cause many genes to be broken in half so they can't function at all. There's also a high risk for cell death or cancer of the affected cell.

DNA Recombination and Chromosome Crossover

DNA recombination is really necessary in meiosis in order to produce distinctly different offspring. Remember that this happens in the *prophase I* phase of meiosis. Now that you know more about genes and DNA, you might want to know how recombination actually happens. It turns out that there are enzymes that help this process happen smoothly. When this type of recombination happens, it's called *chromosome crossover.*

DNA recombination (but not chromosome crossover) happens in prokaryotes as well. In fact, we know more about how bacteria carry out recombination on a molecular level than we do about what happens in eukaryotes. The process in both types of organisms involves enzymes called *recombinases*. These are enzymes that transfer strands from one place to another in very specific ways.

Genetic recombination happens in meiosis between genes. This is a good idea because if it happened in the middle of a gene, the gene product made by the defective gene wouldn't be functional at all. The idea behind crossing over is that *alleles* for given traits of an organisms are "mixed and matched" among the gametes. It's how we have diversity in organisms that undergo sexual reproduction. It looks like this:

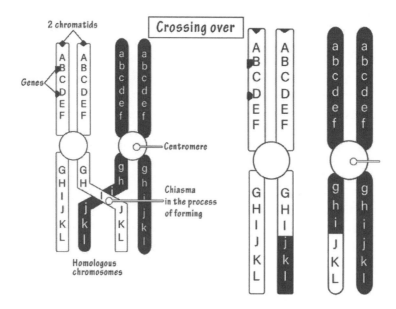

As you can see, the whole sequence requires that the genes line up side-by-side and involves the creation of an intermediate structure called a *chiasm*. This is where the DNA pieces come together and then separate in order to have unique gametes. You can see now why sometimes certain genes get inherited together in ways that wouldn't fit with Mendel's idea that all genes are inherited separately. If two genes are too close to one another, they will cross over together and will be inherited more often together in the same offspring.

In looking at recombination more carefully, you can see how an interesting structure gets formed, called a *Holliday junction*. This is what it looks like:

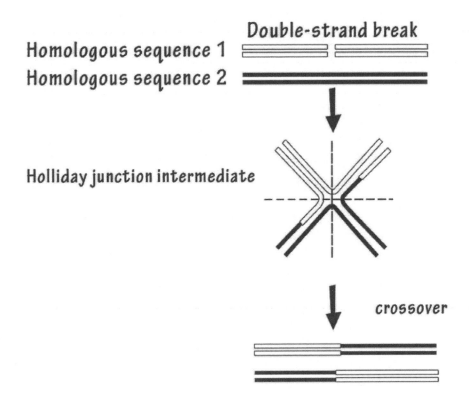

The Holliday junction is the cross in the middle that gets made when the two chromosomes get together and one strand breaks completely. They form the chiasm or Holliday junction and then completely cross over to make the mixed-up pair of homologous chromosomes after that.

To Sum Things Up

DNA gets damaged all the time or it doesn't get replicated properly in the first place. DNA polymerase has its own proofreading mechanisms for repairing any mistakes it has made. In addition, DNA can be damaged by *mutagens*. There are several ways that different mutations can be detected and repaired before they do any real damage.

DNA recombination and *crossing over* are similar processes. Chromosome crossover only really happens in meiosis; it involves the creation of a *Holliday junction* or *chiasm* that allows sections of chromosomes to mix and match as part of genetic diversification of the species.

CHAPTER 37:
GENE REGULATION

After reading this chapter, you'll understand how genes get turned on when they are needed and turned off when they're not. You'll also understand better how cells from the same person can look completely different from one another, even when they share the same *genome*. Let's look first at prokaryotic gene regulation and then how eukaryotic cells regulate their genes.

Gene Regulation in Prokaryotes

As you can imagine, the process of gene regulation in bacteria is simpler than it is in eukaryotic cells. For one thing, a lot of prokaryotes have their genes clustered into related groups called *operons*. This is a great idea because they make all of the same enzymes needed for the entire pathway that's used to make a certain product.

Bacteria use three kinds of molecules in order to regulate their operons. These are called activators, repressors, and *inducers.* Let's see what they do:

- **Activators** — These are molecules that bind to the DNA at a promoter site in order to turn on the operon's transcription processes.
- **Repressors** — These also bind to the DNA but instead bind to separate operator regions on the DNA molecule in order to block the transcription processes.
- **Inducers** — These don't necessarily bind to the DNA molecule and are either made in the cell or are part of the cell's environment. They might turn on or turn off the transcription processes.

Lac Operon Regulation in the Prokaryote

The best-known of these operons is called the *lac operon* in E. coli bacterial organisms. Its job is to make all the genes necessary to regulate lactose production in the cell. It can be turned on or off, depending on the conditions. The goal of the whole operon is to make genes needed for the uptake of lactose in the cell's environment so that it can use it for energy or fuel. This is what it looks like:

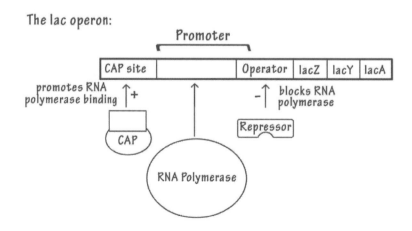

See how there's a promoter region where the RNA polymerase binds? If it is activated, three proteins get made: *lacA, lacY,* and *lacZ.* All three proteins made by these genes are needed to help take up lactose from the cell's environment.

There is an important molecule in the cell called *CAP.* It stands for *catabolite activator protein.* Its job is to help the bacterium scrounge for fuel sources when it doesn't have enough glucose inside the cell. If glucose levels fall inside the cell, CAP gets activated by a well-known second messenger molecule you already know. It's called *cyclic AMP,* or *cAMP.* These get together to make an activator molecule; this binds to the promoter region in order to turn on the operon.

CAP can also act as a repressor in other situations; in that case, it will turn off transcription. Sometimes, the CAP protein binds to the promoter region directly. These would turn off transcription because it does not allow the RNA polymerase molecule to bind properly to that same spot.

Turning on the lac operon with the CAP protein would be a waste of energy here if there wasn't any lactose around to be taken up by the environment in the first place. In order to control the making of unnecessary lac proteins, there is a repressor that prevents CAP from turning on the lac operon. It's called the *lac repressor.* When it binds to the DNA, CAP can't do its job at all. As you can see, the repressor has its own site, called the *operator region.*

Finally, there is an inducer molecule that helps turn on or off the lac operon. It is a molecule called *allolactose,* which comes from the metabolism of lactose. If lactose is available, so is allolactose. This is how the cell knows that lactose is present. In this image, you can see how it binds to the lac repressor, causing it to inactivate the lac repressor. Then, bingo! You've got the operon turned on!

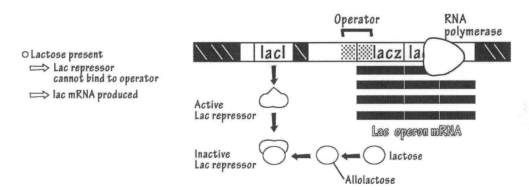

If there is no lactose present at all, there is no need to make any of these lac proteins. This keeps the lac operon turned off continually until the cell "senses" that there is enough lactose to try and get from the environment in the first place. In the absence of lactose, the process looks like this:

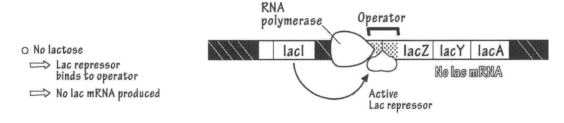

The Trp Operon in Bacteria

The *Trp operon* is actually simpler than the lac operon. There are five genes in this operon. Their job is to make the proteins the cell needs in order to make the amino acid tryptophan. This is why they are clustered together as one operon. If the cell has enough of this amino acid, the operon gets turned off. This is what the operon looks like:

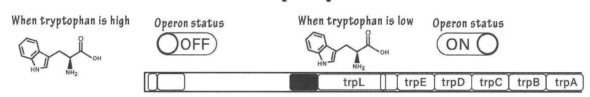

The operator is a stretch of DNA that binds to a protein called the *tryptophan repressor protein.* If it binds to the operator, RNA polymerase can't bind to the promoter region, so nothing happens. This is good for the cell because it doesn't really need tryptophan. If the levels are low, the operon gets transcribed.

This repressor protein needs two tryptophan molecules in order to be active. If there aren't enough of these amino acids around, the repressor can't function. This is another picture of what this looks like:

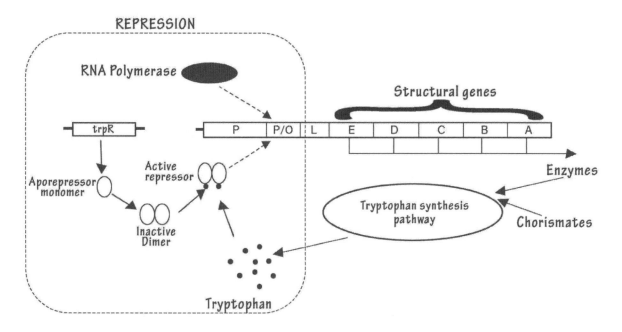

There is a separate gene called the *trpR gene*. It makes this repressor molecule. It is completely inactive without tryptophan attached to it. When they attach, the molecule binds and stops the RNA polymerase from catching hold in order to start transcribing the operon. Without tryptophan, the repressor protein drops and the whole thing takes off, transcribing the genes the cell needs to make more tryptophan.

Gene Regulation in Eukaryotes

Eukaryotes also have gene regulation but it is more complicated than in prokaryotes. There are no operons, so all the genes for a specific pathway are separate.

There are three ways that eukaryotes regulate their genes:

1. Regulatory proteins that bind a long way away from the actual gene they regulate
2. General transcription factor proteins that work with RNA polymerase
3. Gene regulation by condensing or packing the DNA so it can't be transcribed

Remember that eukaryotic genes have the *TATA* box that signifies where the gene starts and where the different transcription factors bind. There are a few other DNA segments near the gene that will do the same thing as the TATA box. They will also bind to transcription factors.

There are also special *enhancer regions* far away from the gene itself. As we talked about, they aren't necessary for transcription to take place but they do make it happen faster. Because DNA is long and flexible, the enhancer regions can get close enough to the gene they enhance in order to have the proper effect on the gene. This is how this whole thing works:

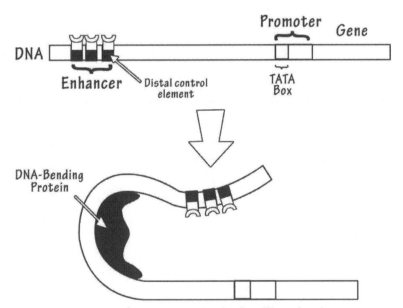

The activators are then positioned to bind to TFs and mediator proteins

We don't know much about all the different transcription factors and enhancers in eukaryotes. The ones we know most about are the *steroid hormones*, which get into the nucleus and bind to areas of the DNA in order to turn on genes.

The Role of Histones in Gene Regulation

Anything that can add or take off a histone protein on a eukaryotic chromosome will automatically regulate the genes on that section of the chromosome. DNA wrapped up in these proteins just can't separate or be transcribed. Most of the DNA of a cell is wrapped up in tight coils because of *histone proteins*, which explains why a cell will act completely different from another cell, even with the same DNA.

It turns out you can make a histone attach to DNA better if you add an *acetyl side chain* onto it. This is called *acetylation*. Any enzyme that can de-acetylate a histone protein can make it attach better to the DNA. This is called *nucleosome remodeling*. It looks like this:

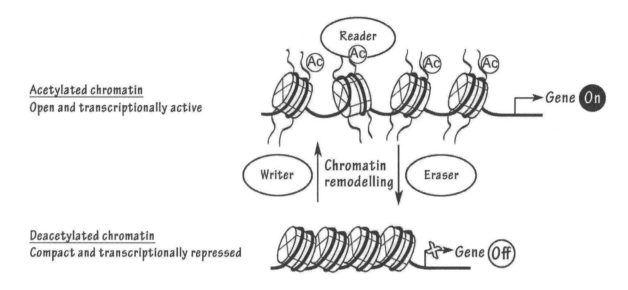

Another thing that happens is that the DNA itself can have methyl groups attached to it. This process will also regulate a gene. Cytosine bases are really prone to getting *methylated.* When this happens, the gene in this area won't get transcribed. There are enzymes that turn off genes just by methylating the cytosine bases on the gene. This is what it looks like, chemically:

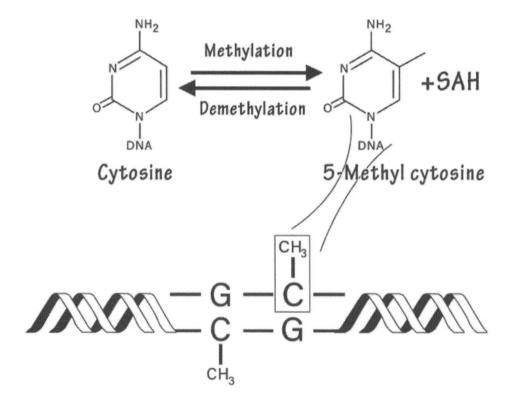

You can imagine that these *methyl* (CH3) groups on the gene would make it difficult for the gene to get separated and transcribed into a piece of messenger RNA, or to make a protein out of that gene's message.

Fun Factoid: *If you get a chance to study epigenetics, you should do that. This is the study of how genes get modified all the time by environmental factors. Events that happen in the womb or in childhood can cause methylation of parts of the baby's DNA, which affects the health of the baby after birth and through adulthood. This often leads to diseases like diabetes and possibly depression. The good news is that methylation isn't permanent. You can change your epigenetics and those of your offspring through lifestyle changes that will reverse the epigenetic changes that happened to you a long time ago.*

To Sum Things Up

Genes have to get regulated somehow. Both prokaryotes and eukaryotes have methods of deciding which genes to turn on and which to repress. Prokaryotes often make proteins in the same biochemical pathway together. These are collected in a group called *operons*. There are *activator* and *repressor proteins* and other molecules that participate in gene regulation in these operons.

Eukaryotes have more complicated ways of regulating their genes. They have *transcription factors* that act a lot like activators in prokaryotes. There are also *enhancers* that act on DNA at sites far from the genes themselves.

Eukaryotes also modify their genes using *histone proteins.* Acetylating these histone proteins will cause them to slip off the DNA so the genes get transcribed, while removing these acetyl groups turn off the genes. Methyl groups can be added to the DNA itself in order to prevent the DNA from getting transcribed.

SECTION ELEVEN:
GENETIC ENGINEERING, OR HOW WE MANIPULATE CELLS

This section should help take your understanding of cell biology into the 21st century. This is where we get into what scientists have done so far and what more is to come now that we know more about how the cell works and how we can potentially manipulate them. You've probably heard about *GMO* (genetically modified organisms) when it comes to plants and animals. But what about GMO humans? Is this in our future as we learn how to manipulate single genes and clone all kinds of organisms?

Let's find out.

CHAPTER 38:
GENETIC ENGINEERING OF PLANTS

We have genetically modified organisms for thousands of years. This all happened when early agriculturists chose which plants to propagate for their crops and which to let die off. It also happened when farmers and dog enthusiasts crossbred animals in order to create some kind of "ideal" animal.

Twenty-first century scientists have taken all of this one step further by introducing genes and crossbreeding in ways we hadn't thought of doing before (or didn't have the technology for). The goal of this kind of work is to genetically modify organisms in just one generation rather than the several generations it takes for traditional techniques used in the past. Let's see how it works.

Recombinant DNA Technology

Most plants are modified using *recombinant DNA technology*. These processes take advantage of recombination in order to create new organisms. With this technology, new genes are inserted into germ cells in order to cause the offspring of these germ cells to have brand new traits that would take a very long time (if ever) to occur naturally.

Why would scientists and agriculturists want to do that? Well, in this age of climate change and the need to feed the planet, wouldn't you want a plant that better resisted temperature extremes, changes in environmental salt content, biotic stressors (like insects, fungi, and bacteria), and drought? What about having crops that were more nutritious than the original species? Or plants engineered to make biofuels or medicines? Sounds good, right?

Recombinant DNA technology can be executed in several ways, but here is one good way to do it in a nutshell:

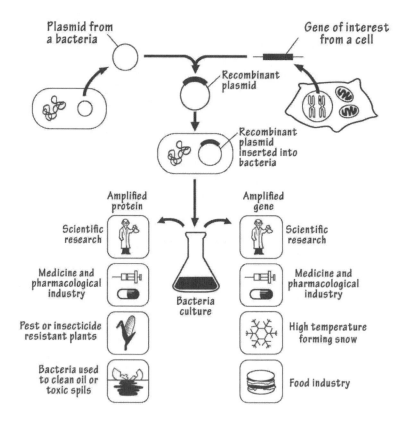

You basically take a gene and introduce it into an infectious bacterium. You can do this in several ways, and the most common way is to take a *bacteriophage* (that infects only bacteria) and put a desired gene into it. Infect a bacterium with the virus and collect the bacteria that have the gene in them, usually in a small piece of circular DNA in the cell (a *plasmid*). Then, infect the organism with the bacteria you've created and collect the new organisms that have acquired the desired gene.

Here are the simplified steps:

1. Find a plant that has the characteristics you want to see in other plants.
2. Find the gene in the plant that causes the desired trait. You can also "make" your own desired gene if you can find an example of in the genetic library (which is a library of all known genes that have been found).
3. Collect the gene using restriction enzymes that cut up DNA; then, isolate the gene you want.
4. Apply *PCR* (polymerase chain reaction) technology, which is designed to take a tiny amount of nucleic acid and turn it into a lot of the same nucleic acid piece.
5. Put the gene onto a *plasmid.* Put an antibiotic-resistant gene into the same plasmid so you can collect bacteria that later take up the plasmid by dosing the whole culture with the antibiotic. Only those with the plasmid will survive.
6. Add the gene to the target plant by infecting the plant with the bacterium that has the plasmid in it.
7. Collect the plant cells that got infected and now have the gene you want to see in the plant.

8. Grow the plant in larger numbers and test to see if it does what you want it to do now.

The most common technique involves a particular bacterium called the *Agrobacterium species.* It's a good choice because it transfers DNA easily into plants. Another interesting technique is the "gene gun," which shoots pieces of DNA onto microscopic particles in the plant cells, collecting those that got "shot." This what gene shooting looks like:

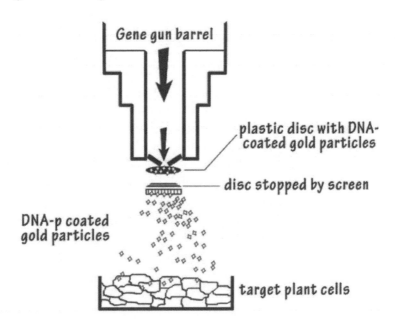

One the cells have been shot, they are grown in tissue cultures in order to make a lot of them that can grow into whole plants. The main plants we are exposed to that have been genetically modified in some way are corn and soybeans, but others, like cotton, rice, and canola, are also commonly genetically modified as well.

How do we apply this kind of technology to the world at large? Here are some potential benefits:

- **Feeding the world more cheaply** — Most of the world is fed on less than a dollar a day per person. Using crops that don't need a lot of fertilizer or pesticides (because they are now resistant to them after recombinant techniques have been used) means cheaper food.
- **Feeding good food to the world** — While the nutritional content of crops might not be that important in developed countries, it's really important where food is scarce. People in developing countries don't have a variety of foods to eat, so the food they do eat needs to be as nutritious as possible. This can be done using GMO plants and crops. Remember, 95 percent of the world's population lives in some developing country.
- **Preventing crop losses** — Crops are lost all the time to insects, parasites, and pathogens. GMO papaya, for example, can now resist pathogens that killed lots of them before. The same is true for GMO corn. This leads to better crop yields and fewer crop catastrophes.

The world of GMO food is highly regulated. The risks have been extensively researched. The outcome of this research? GMO food is not dangerous to eat. Why would it be? You are just eating pieces of DNA in a plant that weren't there before. DNA is DNA. There isn't "edible" DNA and "inedible" DNA. Those who do this work are careful to make sure that only the gene products they want are in the end product. This enhances the safety of the food. GMO food also doesn't cause any more allergies than unmodified food.

Besides using recombinant DNA technology for crops and food, agriculturists and others have used these techniques in the timber, paper, and chemical industries. *Biofuels* are made from GMO plants. Plants are also being looked at as great sources for medicines that can be harvested from crops. Even vaccines and antibodies can be put into a plant and grown … in theory, at least. These possibilities are being studied extensively by researchers.

Could bad things happen with GMO plants? Maybe. Plants like to hybridize with other, closely related plants. If this happens and some type of invasive plant is created, it could create a weed that would overgrow and damage the existing native plants in the environment. Scientists try to get around this by physically separating GMO plants from natural plants.

Fun Factoid: Ever heard of kudzu? While it isn't genetically modified, it is a good example of how an invasive species can wreak havoc on the environment. It's a vine that originated in Japan but was introduced into the U.S. in 1876. Farmers found it was great for preventing soil erosion, so they grew acres of it. Millions of acres later, it started to invade the native species and spread at a rate of more than 150,000 acres per year in some years. It's a good example of why you don't want to have a GMO plant turn into an invasive species.

You need to know that a lot of genetic engineering is already being done on a regular basis in the bacterial world. Where diabetics once had to inject bovine (cow) insulin, they now inject human insulin that comes from recombinant (GMO) bacteria. These bacteria are grown easily and the insulin is harvested for humans to use.

This same technology is now used to make growth hormone, reproductive hormones, human albumin proteins, antibodies, vaccine material, and factors used to help hemophiliacs clot their blood much more easily and safely than was ever possible before. Viruses are also made to be GMO so that they can be used in vaccines after being injected into a person, causing immunity to disease.

To Sum Things Up

Genetic engineering is not new; it is just occurring much more rapidly than it used to be. Using *recombinant DNA technology,* scientists can identify or create a gene that confers a desired trait in an organism. By putting the gene on a plasmid and causing a bacterium to be infected with it, researchers can then infect plants with the GMO bacterium in order to get the plant to have the desired characteristic. Genes can also be blasted into the plant cells with a gene gun.

The idea behind having GMO crops is to make crops that resist disease better, grow faster, or make more nutritious food. This can also make crops cheaper in order to feed a growing population of humans and animals on this planet. There are also vast opportunities to grow plants in the future that make medicines and biofuels, which are other ways to advance the health of the world in general.

CHAPTER 39:
USING GENETIC ENGINEERING
IN ANIMALS AND HUMANS

Okay, so maybe you are more convinced that genetically engineering plants and bacteria aren't bad things, despite the bad press they get. Can you say the same thing about GMO animals or even GMO humans? Is there a limit we should decide not to surpass? Should we go as far as genetically modifying the animals we eat but not humans in any way? What about sick humans? Should we help them with genetic engineering but not use the same technology to create "super-humans"? These are things that ethicists in medicine and technology are actively looking at. Let's see how animals and humans might be affected by genetic engineering.

Genetically Modified Animals

Genetic engineering is used in research all the time. Researchers have created GMO mice that have a high risk of cancer so that they can better study the effects of anti-cancer agents in them. Using genetic engineering in other areas, such as for food, is less common but might also have advantages you might not have thought of.

What do animal husbandry experts do to enhance the survival of their herds or to make them fatter and more nutritious for human consumption? Well, mostly they pump them full of antibiotics and hormones. This does help, but then we eat these animals, also getting a mouthful of the same antibiotics and hormones now found in the muscles/meat of these animals. Unlike extra pieces of DNA (that really don't make a chemical difference in your body), hormones and antibiotics you actually eat can be bad for you.

So, what if you instead created a GMO animal that naturally resisted disease better without antibiotics and that was naturally fatter and more nutritious? Wouldn't that be healthier for you now that you wouldn't be eating all of those extra chemicals? This is the goal of the genetic engineering being done today all over the world.

GMO animals are being created that have a better chance of surviving the newborn period, that are resistant to disease, that grow better and faster, and that have more nutritious meat. Cows are being genetically modified to produce more milk or "better" milk. Sheep are being modified to make more wool at a faster rate. Salmon are made to be GMO so they grow faster for us humans to eat. It's not magic; it's genetics applied judiciously to make an animal more useful to humans in general.

There are three kinds of GMO animals being analyzed:

1. Animals we eventually eat
2. Animals called "bioreactors" that make medicines in their bodies
3. Companion animals

The first two are still developing, but there are quite a few in pet stores and in breeder facilities used for companion animals.

How Is This Done?

One way of doing this is through *sperm-mediated transfer*. Sperm are created that have the desired gene in it; these sperm are used to impregnate unmodified female animals. Some of these offspring will have the desired trait. It looks a lot like this image:

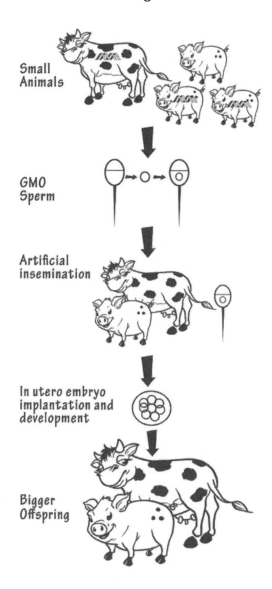

Other things that can be done are using animals to grow medicines. These animals are called *bioreactors* because they have metabolic pathways in each cell more similar to humans than microbes like bacteria. Pigs, for example, can be modified to make pig valves for human hearts that are less likely to cause the person to reject the transplanted valve.

There are also other applications currently being studied. What if we could develop mice that would actually "grow" an ear to be transplanted later onto a human who needs one? This is being considered, and these animals have been created in laboratories:

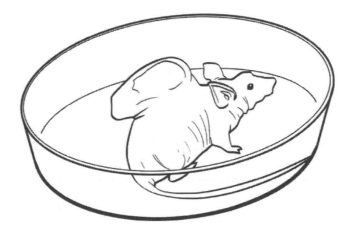

Besides using GMO sperm transfer, researchers are using other ways of making a GMO animal. Some of these include:

- Deleting or adding specific genes into germ cells or regular somatic cells
- Using retroviruses to insert a gene directly into a cell
- DNA injection of genes into the nucleus of a cell
- Transferring an entire nucleus of one cell into another cell

When sperm injection is used, the first generation doesn't fare well, with only one to three percent of inseminations actually leading to live offspring. This actually gets better as you breed the next generations, so that eventually you'll have the desired animal and the birthrate to make it happen on a regular basis.

Simply injecting a "good" gene into a cell isn't really that simple. Remember that genes need things like *promoter sequences* and *terminator sequences* so that the RNA polymerase knows where and when to start and stop the transcription process. These must be added to any gene before it can be put into a cell and made to be operational as a complete gene.

One easy way to do this is to use the promoter and terminator sequences commonly seen in animals that make milk. If you can add these sequences to the gene, you can get the animal to actually secrete a medicine or other agent in the animal's milk.

You can then harvest the agent from the milk. You can do the same thing for egg-layers like chickens and harvest the gene product from the chicken eggs. It looks like this:

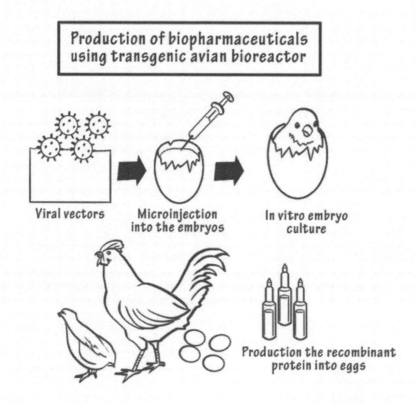

Genius, isn't it? You just have to put a gene into a virus and infect a chicken embryo. Grow the embryo into a chicken. This chicken will then make the drug in its eggs. Grab the eggs and purify the drug, and you've got a nearly endless supply of the drug as long as the chicken strain is kept going.

If you want to create a pig that is less likely to cause a transplant rejection when you put its valve into a human needing a heart valve, you just have to make a baby pig that has some human proteins on the cell surface. The major downside of this is that the transgenic animal usually doesn't produce as many offspring as a normal animal.

They haven't done as much yet to make GMO animals that make meat or other products we buy at the store and eat. Ideally, you might want an animal that grows bigger but that also has leaner meat or that reproduces faster than other animals. Another advantage is to have an animal that would resist disease without using antibiotics. Some think this is unethical to do, which is probably why it hasn't become a popular area of study or research — yet. Sometime in the future, however, it might become important to do this as we find out more about the ill effects of eating animals containing so many hormones or antibiotics.

Cloning Animals

Dolly the Sheep wasn't the first cloned animal but she was probably the most famous. The idea behind cloning is to create an identical baby animal from the DNA or genome of an existing animal. Frogs, cows, sheep, and mice were all cloned before Dolly. Many plants were cloned before these were. Dolly was special because she was the first animal to be cloned from the genetic material harvested from an adult animal. She was born in 1996. Since then, many other similar animals have been created in the same way.

Remember that adult *somatic cells* (non-germ cells) are not made to divide in the same way as embryonic cells. Most of the adult cell's DNA is blocked from being expressed as protein because they can't be transcribed. Only embryonic cells are *totipotent*, which means they can turn into any possible cell in an animal's body. The researchers had to take the nucleus from a differentiated cell and force it to divide like an embryo.

They did it by collecting a nucleus from the udder of a female sheep, injecting it later into the cell of an unfertilized egg cell after first removing the cell's nucleus. They used electricity to force the two parts — nucleus and cell — to fuse together. After they found it was able to divide like a real embryo, they placed it inside another mother sheep so this mother sheep could grow and raise the cloned baby. Dolly was born 148 days later!

How did Dolly do? She was really pampered her whole life at the Roslin Institute where she was studied. She was able to have normal babies (who weren't cloned babies but had her cloned genes in them). At the age of six and a half, she was euthanized — mainly because she had a lung tumor and leg arthritis.

They think that cloned animals don't live as long as other animals because their chromosomes are shorter. Young animals (and humans) have extra sections of DNA at the end of their chromosomes called *telomeres*. These get clipped off as cells divide. When the telomeres get knocked off as the organism ages, the chromosome itself is more prone to damage and the cell dies. This is a big part of the aging of all organisms in general.

GMO Humans?

You might think that no one would "cross the line" to make GMO humans, but, alas, this has already been done in China. In 2018, two baby girls, Lulu and Nana, were born after their embryos were altered to make them resistant to the HIV virus. Interestingly, their genetic father was HIV positive, so there might have been a good reason to do this kind of gene-editing surgery.

How was it done? They used a technology called *CRISPR*. CRISPR is like a pair of gene scissors that can snip DNA in certain places so that other genes can be added. These baby embryos had their CCR5 gene snipped out. This gene is necessary for all cells to allow HIV to infect the cell. Without the gene, these babies could not catch the HIV disease.

While the whole procedure was successful, most medical researchers think this went too far. There is a risk to doing gene editing because it can introduce a bad or mutant gene, or it can remove an essential gene, leading to an increased risk of disease.

No none knows for sure what the future is for this kind of gene editing or for creating GMO or even cloned babies, but there will surely be an uphill ethical battle when and if this gets any further along in medical research than it already has.

To Sum Things Up

There hasn't been as much done on the issue of *transgenic* animals or *GMO animals* as there has been in plants. Still, animals used in transplants, medical research, and in the making of biological products like medicines, have been created as GMO animals that have the characteristics desired for their specific purpose. There are many technologies used for doing this.

Animal cloning is also possible and has been done on many animals. No humans have been cloned but Chinese researchers have been studying GMO humans that resist HIV disease using special technology that hasn't been widely used in other parts of the world.

CHAPTER 40:
WHAT IS GENE THERAPY?

So, maybe gene therapy is the way to treat human genetic diseases. If you could replace a missing or defective gene in a person with a single-gene disorder, you could cure the patient of the disease by essentially vaccinating them with the necessary gene. It turns out that research on this very thing has been going on for a number of decades.

The idea of gene therapy has been around since DNA was first discovered as the source of our genetic material. It's been actively researched and studied for decades. It wasn't until 2012, though, that an actual gene therapy technique was officially approved. This was used to correct disease in people who had lipoprotein lipase deficiency (LPLD), which causes high triglycerides, pancreatitis, and diabetes. The "drug" called *Alipogene tiparvovec* replaced the gene these people were missing.

There are a few considerations necessary before gene therapy for any genetic disease can be effective with the available technology we have today. These include the following:

- The disease must be a single-gene disease that has a known gene.
- The disease should be autosomal recessive, which usually means there is some enzyme missing in the affected person.
- The gene must be able to "know" which cells to go to in order to have an effect.
- The effect of the gene therapy must last a reasonable amount of time.

So, how can it be done? Just as with similar technologies we've discussed, you can use *plasmids* from bacteria, viruses, and *nano-molecules* with genes attached to get the genes into a cell. Let's look at one way that has the best potential of working for more than just a few disorders:

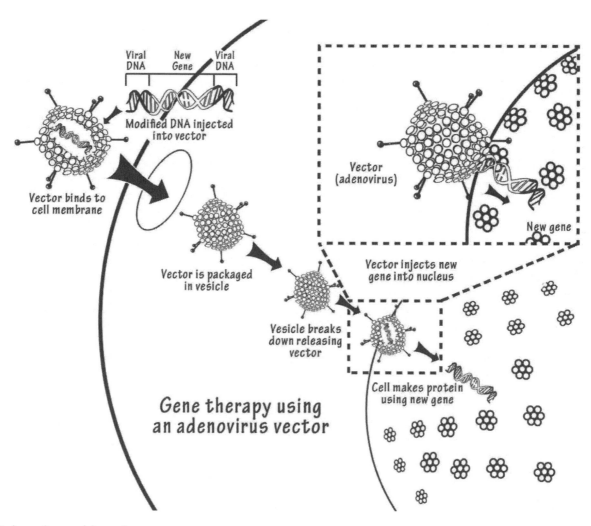

This is how it would work:

1. Get the gene you want and attach it to a viral genome.

2. Put it into an adenovirus, which is a common cold virus.

3. Infect the sick person needing the gene with the virus.

4. Expect that the virus will infect the right cells and will inject the needed gene into the human cell.

5. Hope that the new gene will make the desired enzyme or protein for the sick person.

It was used in the 1990s to prolong the life of some children who had *SCID* (severe combined immunodeficiency disease). The first child to be treated was Ashanti DeSilva, who had a version of SCID caused by a missing enzyme. They put the enzyme into bone marrow stem cells (early dividing cells) and allowed these to grow inside of her. She was essentially cured of the disease. This image shows how they did it:

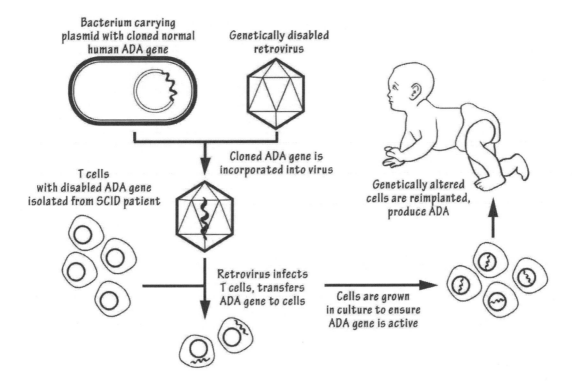

While Ashanti DeSilva is still alive today, not every child treated in the same way survived, so this was a big setback for gene therapy at the time.

In 2015, a boy who had *epidermolysis bullosa* (a rare type of skin disease) was effectively treated using a virus that had the gene he was missing. After infecting some of his skin cells with a virus that had the gene, the healthy skin was cultured and harvested in a laboratory. Then, he had skin grafting of his own "healthy" skin onto 80 percent of his body. This saved his life.

Problems with gene therapy are many and include the following:

- A lot of gene therapies used so far have been temporarily effective. Cells that divide often, like epithelial cells, will slough off, so that the person will need to be treated again to have long-term effectiveness.
- The person receiving the gene will often fight off the vector used to deliver the gene. This makes the therapy less effective.
- Viruses used as vectors can be toxic or cause disease themselves.
- Many diseases involve multiple genes, so these can't be managed with gene therapy as effectively (yet).
- Genes inserted into a genome could enter the genome in the "wrong" spot, disrupting an important gene or causing cancer. This has happened with SCID gene therapy when several treated children got leukemia from the treatment.
- This is a very expensive treatment. The drug called alipogene tiparvovec, for example, costs 1.6 million dollars for each patient treated.

The field could be very large in the future as there are many diseases just like this one. Cystic fibrosis, for example, is perfect for gene therapy but they haven't yet been able to get the gene to stick long enough to have a permanent effect on the affected person. Someday, however, gene therapy will potentially cure all of these *enzyme deficiency diseases* by putting a healthy gene into cells that are missing the enzyme.

To Sum Things Up

Gene therapy is possible and has been effectively used to treat diseases where enzymes are missing in the patient. Several diseases have been treated; however, there is a real risk of causing cancer or having some other ill effect from receiving the necessary gene through the technology we have available to us today.

CONGRATULATIONS!

Hopefully you feel like you fully understand the "language" of molecular and cell biology. You first needed to learn the alphabet — things like atoms, molecules, and other basics. Then, you needed the "words" by studying *macromolecules* and how they interact with one another. The "grammar" involved studying the chemistry of these molecules and how they interact using thermodynamics and energy principles.

Finally, you learned the true language of cell biology — how DNA makes RNA and how RNA makes proteins. You studied cell structures and how cells become tissues or colonies. You learned how cells divide to create the next generation.

The last section was all about how complex this "language" can get by looking at technologies used today and prospects for the future as cell biology expands into the 21st century. Go forth and make use of all you've learned!

Made in the USA
Monee, IL
18 May 2023

34044756R00181